HIMSELF
& OTHER
ANIMALS

HIMSELF
& OTHER
ANIMALS

A Portrait of
Gerald Durrell

David Hughes

HUTCHINSON
LONDON

1 3 5 7 9 10 8 6 4 2

This edition first published in 1997 by Hutchinson

Random House (UK) Limited
20 Vauxhall Bridge Road, London SW1V 2SA

Random House Australia (Pty) Limited
20 Alfred Street, Milsons Point, Sydney
New South Wales 2061, Australia

Random House New Zealand Limited
18 Poland Road, Glenfield, Auckland 10, New Zealand

Random House South Africa (Pty) Limited
Endulini, 5A Jubilee Road, Parktown 2193, South Africa

A CiP record for this book is available from the British Library

Papers used by Random House UK Limited are natural,
recyclable products made from wood grown in sustainable forests.
The manufacturing processes conform to the environmental
regulations of the country of origin.

ISBN 0 09 180167 2

Phototypeset in Bembo by Intype London Ltd

Printed and bound in Great Britain by
Mackays of Chatham PLC

'We'd better ask Himself' –
Member of staff at Jersey Zoo overheard in
marmoset complex at moment of crisis

'It's him himself!' –
Member of public at Jersey Zoo overheard on
sighting Durrell in moment of elation

For Anna Rose and Merlin
in memory of a good friend
and great man

Prologue

ONE MORNING in 1956 I was told, by one of the personable secretaries at the publishers where I worked, that the author of *My Family and Other Animals*, who they had decided was a lady-killer, was coming into the office at noon. One of my jobs as an editor at Rupert Hart-Davis was to alter considerably the work of men of action who could hardly string two sentences together but sold by the thousand. I used to sit gazing vacantly over Soho Square, which in summer was full of girls and trees and drunks all suggestive of freedom, bringing up to scratch other people's books when all I wanted was to be at home plotting my own. This gave me a lofty attitude to the idea of flamboyant explorers deigning to visit the office, flirt like mad with the secretaries and patronise everyone else.

Even to me, though, the promised arrival of Gerald Durrell was an event because his new typescript, *Encounters with Animals*, had just landed on my desk. I had never read him before. My predecessor in the job had left me with the complacent myth, as Durrell was now and then economical with grammar and dropped commas in the rush to communicate his adventures, that I would have to rewrite his work under pressure, expect no credit and receive little reward other than my pittance of a salary. After an interval the printer's proof was then to be sent to Durrell without enclosing his original text. Thus, legend had it, he would never guess that an anonymous wage-slave had guaranteed him yet another best seller, despite his illiterate efforts to wreck his own market.

A glance at the opening pages of *Encounters with Animals* disposed of this legend: there wasn't a howler in sight. But a flip through the script told me that Durrell seemed to enjoy places, animals, landscapes, jokes, wine, weather and people, in roughly any order. Since all these things really did suggest freedom, I thought that Durrell might be the kind of wild-eyed enthusiast who would at once intimidate me by his brilliance.

3

At this point, unannounced, he entered my room. Someone had told him who was interfering with his punctuation. He looked young, blond, handsome, excited and nice, all to excess. After a few words, minding his every comma, he gave me a hard blue stare of amusement and said, 'We're all going to lunch at Bertorelli's, if you'd care to join us . . .' All? That 'all' was crucial. I had learned in two seconds that this man was uniquely inclusive. I had discovered Durrell's gift of drawing others irresistibly into his private orbit. His prompt intimacy had found me a friend for life.

Over plates of pasta he assumed me too to be an adventurer, if not a lady-killer, as well as a very literary fellow well above the sort of tripe he wrote. He and I were of the same generation; I was five years younger. He had only just had success with *My Family* and was regarding it as a fluke. I have by me a proof copy. 'Probable publication date October 1956,' says a white label pasted to the powder-blue binding, 'probable price 18/-.' Meanwhile, as he wolfed his veal, Gerry was working me up with visions of Greece, regretting that by some oversight I had missed that country and therefore most of civilisation, but convinced I would very soon be ordering pints of ouzo with my arm round a willing nymph. Every word was infectious. He had the gift of being as interested in my life as in his own, and as keen to enrich it. As of this lunch I began to see that life of mine in a jollier perspective, riddled with possibilities, ambushed by the unpredictable. It was clear that Durrell had no vanity (and certainly no conceit), only exuberance, a pleasure in the feel of this piece of bread or that plate, the taste of a second bottle of wine, the look of the decorations in this old favourite among dining rooms, the sound of familiar waitresses making a fuss of him, the mood of the moment. The last thing he wanted or needed to talk about, or impose on anyone, was himself, because – as emerged in his most famous book – he so distinctly was himself without trying. That defined charisma.

How and why was *My Family and Other Animals* about to become a phenomenon, a delight to the bookseller, a relief to the classroom, a modern classic?

PROLOGUE

There was the charm, to start with. I was getting lashings of it now, as well as of wine, over that first lunch. He had brought off the next to impossible: imbuing his next of kin, Mother, brothers Larry and Leslie, sister Margaret, with an aura of the ordinary that sounded magic. He put magic within everyone's grasp. The tone raised domestic life to the level of epic without compromising its ordinariness or concealing the quarrels or resisting the jokes, bad as well as good. The book released in you a surge of vicarious relief in rushing off to a country in the Mediterranean and not just sunning yourself but getting to know it more thoroughly than the locals. It made you more human. But that was forgetting the animals. Durrell's hilarious but exact writing never assumed that animals were human. Nor were they alien. They were within the human ambit. On the other hand humans were nothing if not animal. This sweep of a family saga put the universe of living creatures together as one, with plenty of sentiment but no trace of the mawkish, bonding us to our animal forebears and cohabitants of this world with an irrepressible humour that originated in the gut, not the head. When reading we saw Gerry's family behaving with instinct, sibling rivalry, running true to form, just as animals did in nature in their astonishing variety of ways.

The book, as we shall later see, always struck a personal note too. The book had the gift, which Gerry's company had, of making you feel yourself. In my case, and in so many others, the book brought back childhood. Was it my own childhood or the one I never quite had? At the age of nine I was evacuated from London with no sense of wildlife beyond our cat and the mice it hunted. Now I was out in the wilds. In Chichester's surround of dead quiet countryside during the first spring of the war I started hesitantly exploring nature. It was never in Durrell's absorbed way. I was always deflected by boyish lures like biking or bus tickets. But I remember the chilling thrill of hedges when scared birds screamed out, scents of flowers in the heat, the threat posed by insects. The hostile authority of nature gave me goose-flesh.

That lunch at Bertorelli's was the first of countless meals.

And meals counted. I quickly saw that lunch and/or dinner, daily excuses to celebrate being alive, essential punctuation in the busy sentence of a Durrell day, must be for preference long and never less than merry. The menu hardly mattered, provided it was the best of its kind. Fish grilled over driftwood on a Camargue beach, sausages and fried eggs in the bright flat above his zoo in Jersey, the juiciest game in high spots of European gastronomy, all sorts of salads under cherry-red parasols in France, lobsters on a quay at Cherbourg, ramparts of prawns, oysters by the dozen, heaps of buttered bread, an infinite variety of cheeses, fruit – all such meals over the years had been accompanied by staccato bursts of belly laughter which suggested, contrary to reason, that the world was a matter for rejoicing, despite the sombre fact that nobody, realising the desperate plight of that world, could make the smallest improvement to it. With Durrell, over a good dinner, even the horrors became a joke, only to be borne with humour or digested with the help of refreshment in quantity. He appeared to believe, with so little done and so much to do, that self-indulgence alone offered any protection against despair.

The truth for Durrell, as we shall see, was not quite so funny as he pretended to make it.

Soon after that first lunch I left full-time employment as an editor. I worked from my flat as a publisher's reader. My time was my own. I was free except on Wednesday afternoons when I visited the office, the nice secretaries, the slush pile. The spurs to that freedom, as once glimpsed in Soho Square, were sharpened by Durrell's seemingly easy conspiracy with the wicked world. So I hared off into my own future. I wrote a novel that freed me from the ogre of being unpublished and a study of J. B. Priestley whose insistence on being his own man also liberated me. Within a year I was married, happily to a woman with her own life. I moved out of a damp flat in Bayswater, she out of a dry one in Kensington, and we joined forces in an abandoned vicarage in Hampshire which we shared for twelve years.

During that spell I took it for granted that Gerry was a close friend, however little I saw of him. He was always on expeditions to various continents. We were always making films

6

in various Nordic countries. He was trying to find a site for the Zoo he had always wanted. I was trying to find my own voice as a novelist. He continued to support himself by writing books about his adventures. I sought a supportive audience for my inner thoughts and earned a living by other means. We kept in touch at intervals hard to predict. The main thing was that I knew he was there, wherever he was. Now and then hasty letters crossed. Postcards came pounding up the drive in Hampshire out of breath, mentioning possible dates to meet and the impossible jungle he was in. Calls from France to say he was coming through England on the way to Jersey got aborted, were repeated, attempts at ringing back never worked, but days later Gerry turned up all the same, and a summer evening on the back terrace over-looking the downs dissolved into hopeless laughter at the state of the world, our own careers, cosmic deterioration, the oddity of Provence.

I kept being surprised. Lolling on the terrace, whisky in hand, he was cannier than his image as a public entertainer suggested. People cast him as a rib-tickling ringmaster of the comical capers of animals, whereas the core of his life and work, as zoologist and writer, was his complex reaction, one moment furiously hopeful, the next in hopeless fury, to simple facts. These facts had made him bother. They had also nearly cracked him up. The facts of the world disturbed his nights. His days were nagged by them. He had drunk too much on account of those facts. They had helped to cause a breakdown in his mid-forties. Facts had blackened his jokes.

All this came out as over the years, at odd intervals, we talked. I remember his steely blue stare when the hopeless fate of a particular species struck him late on a moonlit night as we sat outside. We had been immersed in drink and laughter. He abruptly sobered. He thrust his thumbnail into my face. 'It's that size,' he said, 'and it's as big in the scheme of things as you or me.' And sat back, grimly gazing at the stars, as if confronting an irreversible fact − which was somehow my fault.

The facts. Over several years I came to know, respect and worry about the facts as Durrell single-mindedly saw them. I had only to see his crabbed handwriting to think, yes, we the human race have ruined the earth, and our own chances, by greedy exploitation. If Gerry looked benignly at a young family slowed down by pushchairs struggling round the Zoo, I thought to myself, yes, our numbers are running riot. The funny little café the Durrells used to have as an adjunct to the manor house where they lived only made me think, well, food supplies cannot last much longer.

I now see that Gerry well and truly educated me, over this long period of our intermittent meetings. He had only to sound off about politicians – 'When they see a gorilla they think they're looking in a mirror and that's an insult to the gorilla' – to convince me, long before it was chic to denounce our elected leaders, that politics of any hue were blind and temporary in their disregard for the battles Gerry fought. That our physical health was in daily danger, he said crossly, was proved by a pollution that was almost beyond control. That our mental stability was in question, he added in rising temper, was confirmed by the fact that none of the money-grubbing conglomerates was prepared to make human or even animal survival the priority. At every level, he observed with fury, the quality of life was on the downturn. It could not be expected to improve, he cried, without an inner revolution, all of us getting our fingers out and starting to think. Nuclear war, he shouted, was a minimal risk compared with the threat of ultimate disaster endlessly creeping closer. Durrell leaned forward to gaze into my eyes with the utmost ferocity. 'It's almost, but not quite, too late to take steps – but what steps? – to retrieve the situation,' he whispered. 'Let's have another drink.'

These issues ate at Durrell. They had always consumed him. But the light touch of his temperament kept him out of the pulpit. He regarded sermons as stones, indeed millstones, prefer-ring to clothe his worry in a motley of jokes and so pass the anxiety on by means of amusement. Whenever we met – first in Hampshire, now and then in London, usually at Bentley's for a spread of oysters and another flutter of fond waitresses who were keeping pace with his ageing and making cracks about decrepitude as they poured hollandaise over the turbot, then more and more in France, when he had eventually persuaded his brother Larry to let him rent or borrow the fairly basic old bungalow in the hills north of Nîmes – I felt forced again to recognise him as the only man I knew who confronted our nasty times head on with humour. Artists being self-absorbed and not coming clean, writers being evasive for want of a message, statesmen employing a host of words to encapsulate nothing, churchmen mouthing perplexity in the cathedrals – Gerry had no truck with all this uncertainty, indecision, flab. His view of the current universe had energy, wit, sorrow, and was simply expressed, often with rage. Most of the rest of us, with roughly his suspicions about the way the world was going, succumbed to the horror. We shrivelled into inactivity or dissolved in tears.

But Durrell's spirit functioned on discontent with himself. He deplored how little he was managing to achieve. By any standard but his own that was lots. Yet he had to keep redoubling his efforts. He survived on his nerves.

In 1961, five years after I first met him, Gerry gave me an introduction to his brother. Lawrence Durrell had recently published *Clea*, the final volume of the *Alexandria Quartet*. His explosion into fame had hardly changed his ways, Gerry assured me, except to improve his brand of plonk. He lived with his third wife Claude in the little house north of Nîmes which Gerry was later to buy. That May my wife and I were driving to the Camargue to make a documentary about the Gypsy festival at Les Saintes-Maries-de-la-Mer and, hearing of this plan, Gerry judged it little short of scandalous not to take some drinks off Larry, as he put it, and beard the ogre in his den of culture.

On the phone from an unreliable café I half heard Lawrence Durrell inviting us to lunch in a voice like a strained and slightly metallic version of his brother's, and on an envelope I wrote down directions which looked so bizarre as to be surely wrong. Two kilometres after passing the French Army, opposite a building that isn't a chapel, meet me at the *guinguette* on the road north to Uzès, he said airily as if reciting free verse, but sure enough there he was outside the roadside bar, summarising us

9

through the window of his beaten-up van, looking eerily like Gerry trimmed and intellectualised.

Until then I had never understood what lunch meant in the Midi. It meant first that it was not served for a long time. This was to leave plenty of time for drinks, often pastis, sometimes whisky, usually wine. The lunch got under way at roughly the moment when people at home would be thinking about having tea. It finished at drinks time, shortly before sundown. By then a friendship had been forged. My wife and I drove back to our caravan among the Gypsies on the seashore along a plane-lined road punctuated at horribly brief intervals by crosses marking the spots where accidents had ended in death. Neither of us was fit to drive. Only I had the gall to pretend to sobriety, remarking on Larry's story of the commercial traveller at Pont des Charrettes who downed fifteen glasses of pastis before lunch, then conducted business as usual in the afternoon. The rule in France was that you were never drunk except shortly after midday. Otherwise you were failing to confront life and ended up with nervous depression. The siesta was invented to allow for these rituals of excess.

Worse was to follow. It began to seem risky to get caught up with these Durrells, especially in France. At the next lunch, a welcome relief from being pushed around by Gypsies who superstitiously viewed a camera as a curse, Henry Miller was present with a girl friend as sensual as one of his novels. Like most lunches that give Bacchus a bad name, it boiled down to the assembled brains talking guff punctuated by one or two start-ling insights forgotten as soon as uttered. I recall nothing either master said. No gem of erotic revelation penetrated my memory. Thousands of decreasingly well-chosen words were delivered at speed over that lunch. Whole books were churned out over the wine, masterpieces extemporised. Culture was reshaped and given new thrust. Larry and Henry were perfectly matched minds com-municating at close quarters across the gulf of the wine-dark Atlantic, Miller mild and gruff and rangily philosophical, Durrell quick, witty, a snappy fantasist.

Meanwhile, hardly speaking, only beaming, our

cameraman eyed Miller's companion and waited to pounce. He bided his time until everyone was too sloshed to notice. His victim also waited, making the odd companionable remark to avert suspicion. It was only when we climbed into our rattly old Land-Rover, having offered Henry Miller and his consort a lift back to the Imperator in Nîmes, that the action started. My wife and I and Henry sat in the front like an upright threesome from some page of *Tropic of Cancer*. Our cameraman sat, no, reclined, no, lay in the iron-hard back of the vehicle with the luscious flesh he had been ogling all day.

I was driving. The rear-view mirror was focused mostly on the road, only partially into the dark back. Jealously I saw movement, or was it an approaching car or a sheepfold receding into the night? I heard a half-grunt. Oblivious to any of these suggestive tributes to his work, Henry droned politely on about the far from intoxicating condition of culture in the United States or indeed here in France, if you only looked around you. I dreaded looking around me. His books had always hammered home sex as the driving force of whatever culture we had, and I was worried about his reaction to the heavy breathing behind his back as our all-purpose crock bumped towards the centre of Nîmes. The bald old coot of American letters clambered out in front of the hotel, shook the hand that had just cuckolded him, and, wearing an unlikely trilby, entered the hostelry that had Roman remains in its courtyard.

It was a night to remember, indeed hard to forget – largely because it bore all the improbable hallmarks of an anecdote told by a Durrell.

Back in Britain in 1970 we moved to Southwark. Putting up the books on high shelves made me look at the Durrells again, halfway up a ladder, slotting the classics into their new home. With an access of pleasure I saw what I had forgotten, that Gerry's thumbprint was on every sentence. You had only to open a volume at random to hear his voice. This voice was often busy exaggerating. It often produced a comparison that defied belief. But, like it or not, you had him instantly in the room with you. The informality of the man writing them made all Durrell's books immediate. On that ladder he caught me yet again by being transparently nice. I wished he were here in this room full of disordered books off the Borough High Street. He was. He was open and ingenuous. He was funny. It was easy to identify with him, whether picnicking in the olive groves of Corfu as a child, eloping from Manchester, persuading the authorities to let him impose a zoo on Jersey, or being bitten by a snake in Bafut. His great gift was to swamp the page with himself without being pushy. He cast aside inhibition. The moment something he was describing threatened to be tragic, he detected how comic it was. He exploded folly, mocked pretension, teased ignorance, nudged error, and continuously insisted there was no point in life unless we laughed at it while being deadly serious at its expense. Durrell's ambiguities were legion. His inner contradictions were sheer pleasure to his friends as well as sheer worry. They were also no doubt a feast for the analysts he was always careful to avoid.

I sat back in our new place in Southwark wishing I could tell a story like Gerry, not to mention press a point with so light a touch and so sure a knowledge that I was right. It was quite a while since I had last seen the old fellow; but by then I knew, largely from knowing him, that paths need not cross all that often for friendship to take itself for granted and gather strength from absence. In any case fate or folly decided that our parallels would join. Within a year, even before putting our London life in order, my wife and I had suddenly fallen under the spell of a remote and ruined house in southern France, within twenty miles of the house where we first met Larry in 1961. Gerry had now taken the *mazet* over. He was our neighbour.

Gerry kept his tongue in his cheek over our efforts to establish a self-supporting commune of two down in that ravishing valley which the twentieth century had left far behind. He raised a slight eyebrow when I told him we were bringing down a pair of Anglo-Nubian goats in a horsebox under export licence from somewhere near Taunton. When I mentioned a classy donkey bred in Sussex which we were exporting via Newhaven/Dieppe he maintained a judicious silence, possibly because it was not a threatened species. He was in favour of our plan to grow

12

courgettes until he heard that we had enough to feed the five thousand but no outlet for selling them. Our cherries ripened only when there was already a glut. He did not think much either of the wine we made in our first year, but never again, from vines planted before the First World War. Even conservative locals, or yokels, recommended replanting. While liking our goat cheese, he thought there was something wrong with our joint psyche for bothering to make it. What was wrong with the market in Nîmes where daily you could get for almost nothing goat cheese, good wine, courgettes and cherries?

We spent four years in France. All we addicts of French life met at the midmorning cafés with yesterday's *Telegraph*, drove miles to drinks with a neighbour in the next *département*, stayed overnight in one another's patched-up mansions, while trying to build a permanent life, trying to come to terms with nature in the raw, aiming for ecstasy. I watched Gerry carefully over these years when I was by no means altogether happy myself. Was he happy? It mattered to me. I knew his blue eyes would go blank if I asked. His gift was for intimacy without question or analysis; his intuition was that friends knew without asking how things were: if bad, having fun helped; when good, you naturally had fun anyway.

I managed to elude paradise after giving it a fair trial. How now to deal with the self-chosen hell? I was alone in London, by an irony house-sitting for a family who that August were holidaying at my home in France. I was elated daily by such stupid novelty as buying a kipper for my breakfast up the street, as if four years in the Midi had been a prison sentence. I drew deep breaths of the bad dusty air of Camden, a northern area of London desert I had never remotely liked. Was this freedom? The weather was hot and tired. For no reason the basement flooded, and I floundered: whom to ask for help, how to mop up? But as if in the grip of some recondite form of nostalgia, I thought of little beyond how I would write my drama about Durrell, whether he would agree – and, most of all, pacing the streets, why I wanted to do it.

Late at night in Highgate over a Greek meal, suitably

enough, I decided why I wanted to do it. So I wrote to him. It
was 22 August 1974. Looking at this letter now, I regret its tone.
But it was right for the time, for me, for Gerry too. It was an
appeal. Out of embarrassment I wish I could cut it.

> A short note to give you a notion of what I had in mind when
> I spoke to you the other day. I think there is a splendid and
> valuable book to be written – which would not clash with
> anything you have done or use material you might be likely to
> need – about you and the Zoo viewed in a detached, sympath-
> etic and informative way from the outside – but, needless to
> say, with devotion. It would contain a good deal of interview
> material with the people around you and with those who visit
> and support the Trust. It would offer a well-rounded portrait
> of you in your daily setting, of a kind which in all conscience
> you would be unable to give yourself. Sooner or later someone
> will be aching to do such a study and I would like the
> someone to be me.

Was I writing a testimonial to myself? Or polishing up my
sycophancy? This was a begging letter.

> I would like to investigate other people's attitudes to you and
> your work, [I went on] as well as to observe the animals,
> and the way you are keeping and breeding them, from a
> layman's point of view. In other words, to write the sort of
> book an ordinary visitor to the Zoo might produce if he had
> the time, not to mention either the privilege of knowing you
> as well as I do or the freshness with which a novice would
> experience the whole enterprise. It seems to me that such an
> account, apart from assisting the work of the Jersey Wildlife
> Preservation Trust by its tone and content, would be welcomed
> by your audience as a different and affectionate view of you –
> I do, after all, see things differently! – and a view over which
> you would obviously have some control. From my own point
> of view the book would represent an exciting piece of explor-

ation into a world which I know nothing about, but of which I recognise the importance.

As for the practical details, I can of course find myself a hovel or attic in the vicinity, and you know me well enough to realise I wouldn't be much in the way – indeed an obtrusive presence is just what the operation doesn't need.

The fact is I had left home without money. After living five years in a backwater heaven, I had got out of southern France for good. I was to pursue a solitary life as a freelance author in London. As with every other big thing in life, my motives for writing to Gerry, or writing about him, were mixed. The motive I least liked had to do with jumping on bandwagons. I would be preying on my friend's bankability, if not edging myself into his limelight. A mitigating factor, if any could be found, was looking forward to a lot of his company. Provided he agreed.

Gerald Durrell lived forty minutes south of the house I had just left in France. Every week or so we had swapped lunches that lasted hours. Now in correspondence he was keen on the idea of a book about him or just pleased to help; I never quite knew which. He seemed to take the sensible view that no publicity for his zoo, his cause, could be bad. But he was not an obvious man. He might be teasing me. He typically implied that even a friend could do him little harm. He had no fear of enemies, only of fools.

At lunch we exchanged collusive glances that promised a lot more lunches. Implicit in the glances was the sober thought that the research would be best conducted over glasses of wine. Only in this way, we agreed in silence, did good books arise from great subjects. Our wives, both present, both to be wives for not much longer, took a more straightforward view. They thought we were conferring on ourselves a licence to print menus.

Durrell, I think, thought me a decent writer who could do better. I considered him a wonderful one who was often slapdash. His advantage was that he always had a theme. Now I could take up his cause in my own way. My instinct was to try to convey something without making any point. All the best messages passed unnoticed. A story was all. In Durrell's books the cause was usually between the lines. Adventure occupied the foreground, fun and festivity the background, and he equated wine with festivity. He also settled his shaky nerves with whisky whenever the going looked hard to a showman who was essentially shy. An ego big enough to get things done didn't mean he was proud of it or confident in himself. Gerry needed his friends. He needed to be surrounded by supporters of his cause. He needed to be made to feel that he had a strong sexual identity and he required of his male friendships a close approach to the physical: bear hugs, mostly, but relations with his favourite men consisted in large part of collusion, if not quite flirtation, certainly of forming a conspiracy for no reason other than it united both parties against the world and excluded women. He was a man whose physical presence was commanding. His weight increased, his girth amplified, but these attributes of self-indulgence only bettered his frame, as the whitening of hair and greying of beard improved his humorous gravity. His eyes of pale but piercing blue and his ebullient voice, and the persuasiveness of his manner (as well as manners), were compelling. I felt agreeably unique in Gerry's company; as did all the friends he invited into his family.

When I was talking to his supporters in Jersey in 1975, I saw in all their eyes the fond, ironic gleam I felt in mine as we spoke of him. Despite being in thrall, we shared our unspoken knowledge that he was overbearing, cruel in wit or tease, generous far beyond a fault, sweet-natured when calm of mind, wholly himself – while the rest of us were just trying to hold together our scraps of personality as best we could.

His life ran alongside mine for over thirty-five years, during which third of a century I almost always knew where he was or what he was up to. Now and then, despite the long gaps, the thought of Gerry's single-minded obstinacy gave me backbone to straighten up my life, exploit it more than I usually dared, be daring. Drink never drowning his sense of purpose, he regarded his animal-collecting trips abroad as a higher form of self-indulgence. He would have to put up with several minor inconveniences. He might suffer a few horribly major ones. But

16

out of any amount of unplanned fun and fuss would come a book. And the chance of discovering, watching, enjoying a rare animal that needed help. And capturing that creature for its own good with a cunning bred in the field. And bringing it home to a circle of admirers, giving it the right food in the right amount of space, and reviving within weeks the millions of years of a species that had flickered almost out of existence. I was envious of such achievement; jealous too. I was always wondering what I was doing that remotely compared in interest or value with any of Gerry's activities.

I knew I could (and would) never write a full-scale biography, detailing his life from the cradle to the present, carefully placing cause before effect. With a subject of fifty alive and well, in mature flow, with plenty to come, such a book would be outdated by whatever surprise Durrell gave us tomorrow. So I settled on an interview spread over several months but compressed on the page into seven days. Every event in the book must happen, or have happened, every quoted word must have been uttered. But for brevity and craft I was at liberty to pack the seven days full of the things he best liked doing; giving parties, travelling in Europe, exploring elsewhere, being at the Zoo, living in the sun. That was the scheme. I thought I might then write a peroration that solemnly furthered the cause. Or just shut up.

I flew to Jersey a few times for a few days each. I joined the Durrells for a spell in a rented house near Grasse which they took to spite Larry when he was being stroppy about the *mazet*; fraternal ructions were just as vigorous in middle age as in the Corfiot days of *My Family*. I saw Gerry on one of his reluctantly flying visits to Bournemouth. I flew over to the Channel Islands with him in a tiny plane that seemed to be stocked with more brandy than aviation fuel. And I spent a warm sunny fortnight with Gerry alone in a borrowed flat in Cros de Cagnes.

Here we were both supposed by our wives to be working hard. His job was to narrate some recent exploring of his own, mine to explore him. Our routine never varied. Gerry rose early, made himself tea, sat on a balcony obliquely overlooking sea and sky, and wrote several pages before I surfaced. 'I wonder if you picked the right trade,' he said with a grin, riffling through his sheets of lined foolscap. He then looked smug and smirked until I told him how many words I had dictated into my machine the previous night after he had gone early to bed. For the rest of the day we regarded ourselves as professionals on an equal footing. We drifted about the apartment drinking, through some temporary quirk of his taste, canned Beaujolais ('It's so disgusting, it cuts your consumption,' he said, forgetting that he opened four cans to every bottle he normally uncorked), while I asked him questions and took notes.

All this was merely a preliminary to lunch. Usually we repaired a dozen paces across the street to a restaurant with a large, shady terrace, a long, inventive menu and no canned wine. The French noon caught us in its inexorably lazy grip. We made it a convention not to talk shop, in other words about himself, during lunch. While smacking our lips over this morsel or that, we let our minds roam indolently in search of witticisms or discussed plans of no great urgency. One of these was where to go in the afternoon for an outing.

We had a rented car which I drove. The country behind Cagnes offered views of and from hill villages, explosions of mimosa in full bloom (or scrambled-egg plant, as we called it), long aromatic greenhouses stuffed with carnations, glimpses of the Mediterranean at many angles from all sorts of eminence. We were in the habit of stopping for a beer where the outlook showed promise or for a pee when necessary. One afternoon, on just such an expedition, we drew up in a forest mostly dark but lit in blazing patches by afternoon sun, and stood breathing resin. On this occasion some curiosity in Durrell stirred, perhaps pedagogy. Zipping his fly with a grunt of satisfaction, he bent down at random and picked up with some care a stone of moderate size out of a bedding of moss and leaf mould on the floor of this not very distinguished bit of roadside nature. Turning it over, he glanced at the underside and said, with the sobriety of tone that was part of his humour, 'Look at that!' I peered, seeing nothing at first, or nothing much, until without further prompting my eyes began to note what his had expertly anticipated seeing under

18

that stone: a society, a series of societies, a warren of tiny holes admitting one species of insect, a larger complex dealing with the size and needs of another, evidence of their foodstuffs, hints of the predatory in minute legs and body shells, the bones or detritus of former occupants of this vast world – and now of course, to benefit my education as well as improve the drama, all was in sudden panic, a universe upset. 'Our mistake,' Durrell said dryly but with pride, moved by the world in his hands, 'is thinking we're not in any way related to other creatures.'

I flew back to London, sorted my notebooks, transcribed the interviews, wished I had never undertaken the project, drew a deep breath, and wrote a portrait of Gerald Durrell in 297 pages. It took some months. By the old low-tech method of stuffing carbons into the typewriter I recall making four copies of diminishing clarity; twenty years ago people never used photocopiers unless they were in public relations. The publisher had one copy. Philip Ziegler has long ago given up publishing to write biographies himself. The agent got one copy, and Peter Grose has moved out of London to pursue his own ends. I kept one myself. And I handed one copy ceremonially to Durrell. A brief silence fell. We met at a crowded party.

'Not you at your best, dear boy,' he said, instantly changing the subject.

There was good reason why the book was no good. I had used it as a means of falling on my feet in England after living too long in France. At last I was doing something useful, attaching myself to a cause. Yet the book lacked commitment. It was written in a lazy spring when I had just met someone new. It was hard to focus on Gerald Durrell's longing to improve a world that to my mind needed no improvement. Life was what was wrong with my book.

Mine was not the only midlife crisis around. By the time my text was complete, Durrell's own marriage had split up. Jacquie Durrell was said to be in Australia, Gerry himself back at the Zoo in Jersey. Friends we knew as couples in France were now singles somewhere else. Expatriate society on the edge of Provence lay in ruins, like most of the villages we lived in. All these accidents of disintegration affected the reception of my typescript, which was intended as a study, though far from hagiographic, of one difficult man's fight towards an ideal with the help of a serene and loyal partner. Through no fault of its own, the book now struck a knell of bathos.

I put the book away. I hid myself from it. I moved from Pimlico to New York. I taught nearly a year in the Midwest at Iowa, then a semester in the deep South at Tuscaloosa, Alabama. Two years passed – memory suggests I spent them commuting on aircraft – then a couple more. I wrote little. The Atlantic became a drone in the ear, a drink in the hand. America from on high began to look like a suburb of something I hadn't yet written. Here far from home I was at odds. Central Park, from the eighth-floor apartment of someone I loved, resembled an urban parody of the paradise I had abandoned in France. Did nothing satisfy me? Into this luxury spread rumours of Lee, a young biologist Gerry had met, was chasing, desperately wanted to marry. She had turned him down. At the furthest extreme of his courtship he had spent hours in a remote Indian post office composing – or rather persuading the clerks to despatch – a telegram of several pages all in one sentence. He regarded it as a masterpiece of long-distance seduction. It seemed to have worked. Or so the story went. Bits and pieces of his success with Lee, his wonder at finding at last a possible partner who was actually qualified with full academic honours, kept filtering through, and then abruptly into the letter box in Manhattan dropped the bombshell of an invitation to their wedding in Memphis, Tennessee, which was unofficially scheduled, as I discovered on the phone, to last five days. Even for Durrell, this seemed excessive. It also seemed all too likely.

In the event it was so exquisitely happy a celebration of a marriage that I remember little of it, except the lawn on which the reception was held, a springy turf composed mostly of wild strawberries, both in flower and in fruit. It seemed that Lee and Gerry were made to be partners, but others will tell that story. Together for two decades they achieved wonders in converting more and more nations, not to mention individuals, to a more positive awareness of the need for wildlife conservation. In harmony they spread the word.

Some months after Durrell died in January 1995, just turned seventy, I was drinking wine in our garden in Kennington with Gerry's official biographer, whom Lee had just appointed. 'Whatever happened,' Douglas Botting suddenly asked, 'to that book you wrote about Gerald Durrell all that time ago?'

I thought for a moment. I had by no means exactly forgotten it, indeed it was a sore at the back of the mind, though well scabbed by now. 'I'll try and dig it out,' I said, wondering where on earth it was and whether I ever wanted to clap eyes on it again. The association was still with a difficult patch in my life, as it was with a bad spell in Gerry's. 'If I use any of it,' Botting said amiably, 'I'll give you full acknowledgements of course.'

My first thought, as a fellow author, was to give Douglas every assistance. My second thought, also as a fellow author, was not to. 'It might just be in our place in Wales,' I said. I detected in myself a reluctance to hand over original material even to a worthy cause. 'But I've really no idea.' I found I wanted to delay finding the typescript, not exactly to hold him up but to curtail his unrestricted use of it.

I poured us more wine. But my conscience grew uneasy. A moral issue of my own making seemed to be nagging me. Then to my rescue sprang the simple question: what would Gerry say? He would have thought me a dunderhead on several counts: for suppressing a story I wanted to share with likeminded people, for passing up the chance of saying what I wanted about someone I admired; but most of all for even contemplating the idea of letting something go for nothing. His was a life built on fund-raising and self-subsidy; it was a principle honestly to squeeze as much money as possible out of anyone who had it, money being utterly unimportant except when you needed to spend some.

I had no memory of my book. For some weeks I put off the expedition to Wales on the selfish ground that I thought I was conducting the search just for Botting. He had met Gerry only once. I had been a friend of Gerry's since our twenties. I

began to feel oddly protective towards Durrell, untypically defensive: and what were my own interests? Or Gerry's? The notion
of seeking out part of my past on paper rather scared me. But
why not have a go?

The search for the missing typescript then took on the
feel of a quest. In a less exotic way it had something in common
with one of Gerry's expeditions. I was off to Wales to try to trap
a handful of pages which might – I had to believe this – hold the
secret to my subject. In those forgotten paragraphs I might have
captured him. They were now a threatened species. The whimsy
of this didn't worry me. In fact, the idea got me into the very
sort of anticipatory mood in which Durrell himself probably set
out on those trips, in a real hope of finding an unrepeatable
something, a missing link. So I got into the car in high fettle,
determined to run to earth those pages. Gerry would have understood it completely, my obstinacy, but also thought it ridiculous.
If you lose a few pages, he would have said, you just write them
again and they'll very likely be better. But these were twenty years
old, prehistoric pages, as lost as a dinosaur. They belonged to
those times. They were irreplaceable, even if they were no good.
Our property in Wales, where I felt almost sure they were in
hiding, was at the back of beyond. The barn where I might find
these pages was at a further remove. No longer doubting for an
instant that the search was worthwhile, with only a day or two
free to find the work, I set off westwards from London.

As always, the road felt as though it were going nowhere
but to the only place that mattered. I was thinking that I liked,
and Gerry would like, the idea of this book being open-ended,
not drawing conclusions, leaving the reader to make up his
mind not only about Gerry, his nature and his concerns, but my
own feelings, strong but mysterious to this day.

In my hayloft in Wales, unvisited for well over half a year,
the smell was of tarred beams and a welcoming damp. The tabletop
was spotted with dried droppings from birds. A robin that lay
dead on the floor was almost weightless in the hand. One or two
wine empties stood about, gravestones of self-indulgence, with
labels I never recalled drinking. The really bad sight, worse than

PROLOGUE

I remembered it, was the tumbled and tangled heaps of papers in cardboard boxes and rubbish bags piled one upon another in a corner of the hayloft, actually occupying a quarter of it to quite a height. This is your life, I thought, winterlong damp already chilling my spine: the records of a misspent youth, an uncertain maturity, a defeated age. Why on earth had I not thrown things away as I went along? Well, it might look sad, this mess, but it could have happiness in it, for all I knew. I reconciled my mind to a long search. In all that chaos, having gone as risibly far as this to rescue it, I recognised Gerry as the only item worth having. Here were my memoirs, attired as garbage, rotting away. The rest had better be silence.

Then I found it.

I started reading. On the first page I smiled involuntarily; and Gerry was back.

Working over the text has made me realise in no sentimental fashion that Gerry is only in the obvious way dead. So here it is, much cut, shortened by a third, crisped up, far better for it. I find myself listening off my own pages to what he has to say. His voice is alive. This is no emotional illusion; or, if so, it hardly matters. I see it more likely as a collusive act with his army of readers, young, old, moving up through the generations, retaining their loyalty, who will always think of Gerry as an immediate presence and, as things grow daily worse in the environment he did his best to serve, now consider him a well-loved pioneer.

Sunday

GERALD DURRELL lay on his back in bed snoring. His belly bulged under the sheet. He had the look of a bearded matron on the verge of going into labour. Outside a shot echoed down the valley. French hunters were on their weekend rampage against what little wildlife survived in the Languedoc. Even so we had been promised local pigeon for lunch.

The rest of us moved about the house in whispers. We had our fix of coffee. The energies of the house would start to function only when his nibs staggered out of bed. Having a minor divinity on the premises kept everyone on tiptoe. Even his snores sounded as peremptory as orders.

The day promised heat. For five years Durrell had spent his summers in this *mazet*, a smallholding lapped by slopes of olives on the northern outskirts of Nîmes. It was no distance to restaurants, airports, garages, supermarkets, his quartet of essentials. Yet it felt remote here, ageless. A dry landscape of scrub, where lizards haunted the rocks between gorse or thyme, was a skeleton key to his memories of Corfu, a guide to the childhood he had made famous.

To Durrell's annoyance, the house still belonged to his brother Lawrence. He let it to Gerry in the high-handed manner of an absentee landlord, yet he was always on the doorstep. Over fifteen years ago much of the *Alexandria Quartet* was written in this whitewashed house. From the terrace, shaded by old almonds, the view was of miniature hillsides on which one or two breeze-block villas were going up. These eyesores constructed by Algerians were regarded by both brothers as a risk to the environment. To the rear of the property's fifteen acres stretched a blasted heath occupied by the military. The writing of *Mountolive* had been accompanied by the chatter of machine-gun fire. Here the elder Durrell had built dry-stone walls and a worldwide reputation. The only thing he now claimed from the property was

27

the yearly harvest of olives. He liked to have his own oil at home in Sommières a few miles away.

The snores drew to a close. A plump hand parted the plastic strips that kept flies out and with sluggish dignity the tall, broad figure emerged on the terrace. Durrell stood a few moments, blue pouched eyes blinking slowly in the morning light. He acknowledged nobody. The set of his head, deep in the shoulders, was both defensive and challenging. The bath towel draped over his belly hung round his knees, a cockeyed skirt vaguely suggestive of hula-hula. He sniffed once, through his bulbous wedge of a nose, then turned back into the house. The sunrise had passed muster. An early bee slipped past at speed. There was little birdsong. Most of the dawn chorus had been mashed into thrush pâté, a speciality of the region.

The shower hissed within. A few tubular complaints arose from the plumbing. An absolute silence fell. Prone on a towel, Durrell was at his morning yoga. The usual stint was half an hour, more if he suspected the day would make excessive demands on him. Since a breakdown in health five years ago, which banished whisky from his intake and taught him to smoke without inhaling, he had so convinced himself he was dependent on yoga that he never missed a chance of persuading others. He rejected bedrooms in continental hotels if they lacked the floor space for his exercises. More conventional gym he brushed aside. 'There's no point in my doing press-ups,' he said, 'when nothing I can do prevents my stomach touching the floor.'

Working from a 50p manual based on a television series, Durrell now moved in slow motion through a varying routine – cobra, lotus, bow, headstand – without subjecting his body to the least strain. 'The beauty of yoga,' he said, 'is that it asks only what you want to give it, like a perfect wife, and it then lives with you all day, unlike a perfect wife. I would make it, instead of football, compulsory in schools.' He gave a faint groan of pleasure as the effect bit, wiping out traces of dream or hangover, turning his face 'boiled-baby pink'. Outside some geese created a racket across the valley and a concrete-mixer coughed twice and began chuntering. Inside Durrell sat on his mat, face blank above the mournful beard, a potbellied Silenus impersonating an inscrutable Buddha.

With slightly puffed ceremony he got up. Hitching the seersucker sarong round his waist, feet splayed in trodden-down espadrilles, he flapped to the kitchen and made some tea. There was plenty of tea around, sixty packets or so. Despite an almost patriotic prejudice in favour of the Midi, the Durrells never bought expensively in France what could be brought easily from Britain – tea, tissues, salt, curry powder. There was a glum pleasure in filling a kettle, watching it boil, spooning tea into a pot: a household task momentarily shielding him from pessimism or nerves. He brought his tray of tea out to the terrace, and sat drinking it in a basket chair canted sideways by his weight, content with the tea's heat and the morning's warmth until something preferable caught his attention.

And soon it did. Under the rough table on the terrace were a number of dry pips from a watermelon eaten yesterday at lunch. They were being carried off by a long train of small ants. One tottering ant was trying to join the column with a plump dead fly, straining every limb. Durrell, who had been brooding over his bare knees, now looked between them. On the crazy paving at his feet life was noticeably going on. His face tensed. Out of the slothful middle-aged tea-drinker emerged a fully fledged infant zoologist. 'You realise, of course,' he said in tones tinged with awe, 'that what those ants are doing is beyond any human power. To carry a seed like that is the equivalent of you bringing an elephant home in your arms, and, as for the fly, it's approximately similar to me dragging a blue whale single-handed out of the water.'

Durrell sat back. 'Ah,' he said, sipping tea. He gazed at the thicket of passion flowers in the tattered bamboo awning overhead. A butterfly hovered into view. 'Ah,' he repeated sharply. 'A skipper. Good. That makes seventeen different species of butterfly I've seen in France. What a very good start to the day!' He watched the butterfly zigzag through blue air, loped into the house and brought back, fingering the pages with relish, a field guide to Europe's butterflies. Within seconds the specimen, still just visible flitting past a concrete-mixer, was pinned down as Foulquier's grizzled skipper. A rarity.

The sky was already a deep, hot blue. Two or three dogs sparked one another into aimless barking across the valley, a flying beetle collided with the awning and tried to enter the hollow of a bamboo. The ants pursued their long trail underfoot, a heavy wasp lurching on to the table whizzed off, an elated twitter of goldfinches faded behind the house. Durrell sat classifying these events in silence. His contentment was obvious but edgy. The sun lifted. Heat penetrated the skin. The cicadas started a moody zizz in the grasses and a tree frog stuttered in one of the almonds. Despite lots I didn't know about him, I felt a lazy reluctance to ask questions in weather like this. Much of my old friend's writing had the apparent innocence of such a morning. His gravity of manner often suggested a child pretending to be grown-up. His enthusiasm for that swiftly identified butterfly, the ants dragging overweight seeds, was infectiously boyish. All his life he had kept this freshness; his interest in nature, in animals, was obsessive from birth. 'I know I've said it somewhere in my books, which of course nobody believes,' he said now, finishing his tea, 'but my first word was, quite literally, "zoo".'

He was born, Gerald Malcolm, in India on 7 January 1925, a Capricorn, a late child of the Empire and by several years the last of the family, Lawrence George (1912), Leslie (1916) and Margaret (1917). His mother's earlier pregnancies had been normal, but to celebrate this last-born she not only grew to three times the anticipated size (which for a tiny woman created such problems with clothes that she refused to be seen in the club) but also succumbed to an irresistible yen for a daily bottle of champagne. 'Instead of a silver spoon,' Durrell said, 'I was born with tinfoil in my mouth.' The usual host of superstitious Indian servants had greeted the other three births with polite indifference. This time wild excitement welcomed the coming arrival. As they served champagne to the minute but monstrous lady under the drawing-room fans, they kept saying that this latest Durrell would be immensely lucky. 'All this of course was hearsay,' he said, 'which probably means that the truth was even more unlikely.'

His first recollection was a response to an animal. For some reason, untypical of so cosseted a childhood, he was left unattended at the age of two in a garden of a house from which the ground sheered steeply away. A horse was tethered on too long a rope. He recognised the horse; he had been walked round in circles on its back. But now the horse strayed close to the precipice, the rope tightening round a leg, and in the slow motion of memory fell sideways, leg snapping, and tumbled over the cliff. It was a primal feeling that still haunted Durrell after a lifetime: identifying with the animal, wanting to help it.

His few years in India yielded no memory of a human to match that of the horse. His parents, comfortably off and devoted to each other, had left little trace in his mind of the pampered routines of well-to-do colonials. Yet, even unremembered, the subcontinent must have formed his tastes and expectations. He was to have a lifelong dependence on sunshine, a blazing sky, vegetation zithering with insects, persons to do his bidding, strong refreshments. Apart from the prenatal presence of champagne in his veins, wine was confirmed as one of life's essentials when, at the age of two, he was prescribed for tropical diarrhoea a daily glass of port.

On the outskirts of the hill town where they lived stood a tiny zoo. It was a hotchpotch of redbrick cages topped by cupolas. The memory outlasted images of his father, who died when Durrell was two. Scruffy tigers padded to and fro behind black bars. The heat made the uncleaned cells stink. 'Yes, yes, it's that smell I remember,' Durrell said, 'and the *feel* of animals, you know, almost a warmth coming out of the cages, like a fire on a cold night.' Whenever his ayah, who took him twice a day for a walk, made the courteous error of consulting him about the itinerary, the boy insisted on returning to the zoo. At any attempt to thwart him, he turned purple with rage. 'Oh, dear!' said his mother, 'Why not just give in to Gerry?'

One day on the way to this zoo to view the steaming excreta of lions, the ayah paused to gossip with a woman in a bright magenta sari. Bored, the boy drifted to the roadside ditch where two large khaki slugs were mating. He crouched over the

31

slugs in baffled fascination, until suddenly he found himself dragged to his feet and told by the outraged nurse to leave them alone. Why? Because. Why, why? Because it was dirty. And up the hill they trailed towards the zoo, lured yet again by the perfume of tiger's urine in the heat.

To amuse the boy whose birth they had greeted with rapture, the servants now showed him how they made animals out of wet clay. It started as a game, but they had underestimated his interest. Every day he demanded a new animal to add to the growing collection, as if creating from scratch a miniature zoo of his own. It was reality becoming art – and art proving just as real.

At this weighty point Durrell paused, gazing into the heat haze intensifying across the valley. 'Oh,' he said, as if surprised, 'I remember eating lamb chops. I suppose that's an animal reminiscence too. They were lamb chops covered with breadcrumbs in a stifling hot train travelling up to the hills . . .'

The one memory of his father, as sharp as a snapshot, had survived thanks only to its animal content. Gerry saw gigantic pillows, his father's head resting distantly against them; and, as though levitated by memory from the scene, Durrell stared down at his own baby body crawling up the miles of bed towards his father, while hearing a voice recount the adventures of the three bears. That was all. Paternal love, or filial, was a fairy story about animals. There lingered in his mind no glimpse of his father's death, no funeral, nobody in tears. Either he had been gently removed from the scene or drawn a self-protective veil over it. From that forgotten moment he became a mother's child. He remained so for the next forty years.

Only yesterday I had talked to Lawrence Durrell about their father. Within the family his death by a stroke in his early forties was said to have been induced by worry. He had amassed a typically colonial fortune building railways. It was when he turned his technical attention to the roads that he lost ground. In one ill-starred venture he undertook the construction of a highway on a fixed-price contract, only to find that the subsoil was solid rock.

Even Larry, thirteen years older than his brother, recalled

little of the character of the man. An itinerant engineer, he lived out of suitcases, rather like an actor. His returns home were intermittent. The family switched from one company house to the next, never domiciled in one place for long, never fully united. To this day the impression of the father was of someone ordinary and sincere, hard-working, hardly sensitive. 'My mother was the neurotic,' Larry had said. 'She provided the hysterical Irish parts of us and also the sensibility that goes with it. She's really to blame for us, I think − she should have been run in years ago.'

Sitting on the terrace in the sun, I mentioned this sally to Durrell. Nothing was confidential between brothers nor, usually, between their friends. 'Did Larry really say that?' he said. He stared at the sky for a moment, smiled dismissively, then struggled out of the basket chair. 'The moment has arrived,' he said, 'to press a tiny glass of red wine to the left kidney.'

In a few moments he returned with a corkscrew, two glasses and a bottle of mildly superior supermarket *ordinaire*. The day's first drinks were poured. The morning expanded. Sounds of activity emerged from the house. Copper brown and wearing a two-piece swimsuit, Jacquie Durrell strolled out to hang the laundry on a line slung between the almond trees. 'Where are we going for lunch?' she called.

There was no reply. Tensely bent at the waist, Durrell was stalking. 'Look,' he said. 'Come and look at these hunting wasps.' Not moving an inch, we stared at the wasps, caught up in their activity as they shifted in and out of the sunlight. Gerry's cupped hand waved vaguely in the air. He had remembered his wine. He stood up, sipped it and said, 'Now where are we going for lunch?'

With asperity Jacquie pointed out that she had already asked. 'I want a delicious and for preference unspeakably expensive meal,' Durrell said. A brief altercation about current prices took place. The *Guide Michelin* was brought to the table. For starters the relative merits of a *mousseline de saumon* at Avignon and a *pâté de canard* at Nîmes were discussed. He smacked his lips over the beef roasted in the baker's oven of that little dark restaurant in Bagnols-sur-Cèze. 'Hm,' said Durrell deeply, as though speedy choice would ruin the meal in advance. A salivating session

33

ensued. The *mousseline* would consume the most petrol while providing an expansive drive through a hungry noon. But the tender beef? Not to mention Le Grau-du-Roi where we could have a dish of those little local shellfish *tellines* washed down by lashings of that rather good dry white Listel. Michelin was carefully closed like a Bible and Jacquie strode towards the pool, grinning over the male self-indulgence and no wiser. 'I'm going for a swim,' she said.

In the pause Durrell again consulted his glass. Bees whipped past, butterflies pursued their dotty course around our ears, cicadas buzzed towards a climax which at noon, when the heat was at its height, would stop dead, as if to allow space for the French to have lunch. Durrell's chest was pouring runnels of sweat on to his stomach. He patted it reminiscently.

'I well remember on the ship coming back from India,' he began in the accents of a high-ranking officer, by imitation puncturing his own pomp. Belching, he stared accusingly at the glass. There was an expectation that someone should refill it. I did so. 'Thank you,' he said staring blankly into the past. The ocean had been rough, the sun hot. All the parents were sweating buckets and retching in their bunks. But all the children, in tiptop health, were crying out for entertainment. And Gerald was suddenly sitting in memory, at the age of three, on the edge of a billiard table with other children, watching fluid lines shifting at great speed on the wall of the darkened saloon, with no idea what they were – until abruptly it clicked: the shape of cats, a cat walking on two feet like a human being, running up a vertical wall unlike a cat, but a cat nonetheless as the blur focused. It was a cat called Felix, all of a sudden comprehensible in a dimension he had never encountered. He was seeing a film.

His first cartoon gave an early suggestion of the cunning of art, and convinced him that animals were everywhere – not just in zoos, or clay models, or breaking their legs in fields, or characters in stories. They were everywhere and for everyone. With the family on its way back to England, the youngest son had animals and movies stamped into his system for good.

Without a father, the family returned home to a country

that was in no sense home. Mother, born in India, had never known England. Possessed of a small fortune, she bought a 'preposterously elegant' house in Dulwich with enough floor space for the carpets, acres of pile handmade in India to her own designs. This was the first of many acts of innocent folly which dissipated the fortune before Durrell was far into his teens. Such carpets demanded the tread of a butler to do them justice. A butler was duly engaged. The boy Durrell found himself in the south London suburbs but also in a grandiose approximation to the Raj. To complete the picture, a dog appeared on the scene, a bull mastiff of commanding size, called Prince.

Prince was London for Gerry. Of brother Leslie, who came home to lunch from Dulwich College, no trace lingered in his memory. No doubt Larry was active in the background, busy concocting his juvenilia, but seemingly without any connection to the family. Two diminutive women, Mother and Cousin Prue, both fond of everyone and timorously lunatic, haunted those years with their chatter. They occupied a world far removed from the boy's solitude where only animals belonged. It was Prince who commanded the foreground, Prince with whom he played for hours in the capacious dining room, riding him bareback, sleeping on him, being treated as a puppy by him, Prince who with a weighty paw held the boy down and licked his face.

But the mastiff's sweet nature was shadowed by his hatred of other animal life, most of all dogs. When taken for walks by the two little ladies, Prince dragged them on his lead to the park, where he picked up poms and pekes in his great jaws, shook and dropped them, until dog owners rushed their pets indoors uttering threats over their shoulders. In the end, by popular demand, Prince was banished to a lonely farm where, Durrell assumed, he spent the rest of his life 'assassinating the poultry'.

At home when Prince was naughty he was locked in the drawing room 'for psychological effect'. The dog protested about this treatment only once. Gerry was in bed upstairs, the drawing room in darkness. Prince stood at the door, refusing to enter, hackles rising, snarling into the gloom. Mother switched on the lights, stared, peered behind curtains, found nothing and eventually decided to wake Gerry for a second opinion. Gazing into the depths of the room, the sleepy boy said that he could see his daddy: seated comfortably in a fireside chair, wearing a favourite suit he donned in the evenings, smoking his pipe.

There was no doubt or hesitation. Though a child aroused from deep sleep, perhaps dreaming about his father, might have projected on to the scene a figure he missed, the dog's snarls confirmed a presence. To that room, months and months after his death on another continent, clung a person who haunted the family on and off for years. Mother missed him enough never to marry again.

A year or two later in reduced circumstances, the family was living in a flat above the Queen's Hotel in Norwood when Durrell's father more than once made his presence felt, though in a fainter guise. That hotel formed the setting for Larry's first novel, *The Black Book*, which he was writing, using his father's rolltop desk. If Larry left the desk open while he was out, the desk was often tightly shut on his return. Larry judged it to be a protest from the grave against his writing erotic filth on a hallowed surface. *The Black Book*, published in Paris in the Thirties, was banned in England for three decades.

Norwood faded into the past. With care Durrell put down his empty glass, glared, as if challenging anyone to doubt these early brushes with the supernatural, and walked with blimpish authority to the swimming pool at the rear of the house. No more than a tank with a rough silver lining, half a dozen breast-strokes one way and none the other, this pool was overhung by olive trees. Equipped with an up-to-date filter (the only improve-ment the Durrells had made to the property), it looked as antiquated as a Roman bath and functioned about as well. Durrell plunged nude into the water, vanished long enough to let the surface settle, came up breathing easily and put his bearded chin on the poolside, which commanded a spectacular view of the valley. A number of wasps, sun-drunk and wet-winged, tumbled unsteadily round his face. With disconsolate pleasure he gazed at terrain which belonged to his brother but by now, after years of camping in the house, really ought to be his.

A retired parson in East Anglia had led the Durrells to the water supply for the pool. This too smacked of the supernatural. A large-scale map of the surroundings of the *mazet*, posted to Suffolk, came back marked with the spot where at 250 feet they would find all the water they needed. Ten feet short of that depth, the experts recommended the Durrells to halt the drilling which had hit only solid rock. But they persisted despite the expense and found water.

Durrell believed it wouldn't take much to turn the valley into a paradise. Irrigate the slopes, plant peach trees between the olives – yes, and nectarines too, apricots, cherries, all the fruits so delectable in this region, eaten straight off the tree with their bloom still warm from the sun, ripe juice spilling into the mouth on hot afternoons – and all thanks to the water. A local couple to live permanently in the other little house, providing the Durrells on their summer visits with eggs, vegetables, even game: they could, for example, raise quails. The very mood of the landscape seemed to change under the visionary force of Gerry's plans for it.

Within twenty minutes we were in the Mercedes on our way to lunch in Avignon. Gerry also owned a Triumph Stag convertible, which was to be sold. It had let the Durrells down with a series of minor breakdowns which had forced him 'to spend more time in bad French garages than in good French restaurants'. He had vowed never to travel in it again, even to buy a newspaper, which in any case he never bought. They contained only lying versions of what he already knew to be false.

Jacquie was at the wheel. She spoke little, except to extol the car as a miracle of Teutonic efficiency. Silence fell as we passed the barracks north of Nîmes and turned towards the Rhône into hill country, where tumbledown villages of faded ochre, unchanged since the First World War, crowned the uplands and gathered silently in the valleys. Now and then humming a snatch of unrecognisable tune, Durrell sat immobile, staring at the unwinding road. When someone commented on the beauty of the distant Alpilles softly etched in a purplish haze against the white heat of the sky, he glanced up briefly and muttered agreement. His present was at a low ebb, much of him dispersed in his past.

Durrell's first zoo began in London with a lead camel when he was five. The foundations for it were laid in a big shiny store, the boy darting between people's legs to see, nose at counter level, the tiny bright-painted creatures arranged desirably in neat platoons. They were all the same, yet all quite individual, in a Woolworth's reservation stuffed with metallic game. At home he collected from the garden a selection of stones to form their habitat, in which wilderness, across the grasslands of the drawing-room carpet, he let them roam at will. To control their tendency to wander and be crushed underfoot, Mother insisted on the construction of zoological gardens in the form of a wooden box which Margaret, then thirteen, promised to furnish with shrubberies made out of twigs. She at once vanished to play tennis, which the boy took as meaning she had gone for good. His bitter complaints roused Leslie to do the job, whereupon Margaret, returning from tennis to find herself denied the pleasure of arranging a zoo, fell upon her elder brother, biting, kicking, punching. Gerry, delighted that his interests had inspired such emotion, sneaked off to alert Mother, who was giving a tea party for several prim ladies. She quelled the squabble with a slap, but the zoo survived. It was the first local difficulty over Gerry's passion for natural history. Later troubles were to give much of the comic bite to *My Family and Other Animals*.

He still saw them all, that camel, the penguin, an elephant, a pair of tigers, all kept strictly apart in the orange-box zoo. They were never given names. They were proper animals. 'No leopard was called Popsy,' Durrell said. 'I had an eye to the zoological niceties.' As endless rearrangement kept the boy quiet for hours on end, Mother decided the collection was her one successful investment. To compensate for the time wasted daily at the infants' school in Norwood, Gerry's concentration at home was total. His imagination was in thrall to these animals. They were more real to him than the so-called realities of his elders. 'It's a wonder Mother didn't think I was mentally retarded,' Durrell said; then, after a pause: 'She probably did.' Another pause. 'I possibly was.'

We arrived in Avignon, a city 'cruel and famous' Larry called it in *Monsieur*, with its 'shady streets and quiet shabby squares'. In today's heat the main Place de l'Horloge mocked that romantic image. Buffeted now and then by teenagers doing their own thing, we strolled through a muddle of youth in well-travelled jeans, waving battered guitars and posters inciting rebellion. Leaflets dramatising a variety of causes were scattered on the pavements where tourist families lunched under parasols.

Durrell passed through it all blankly, his features firmly set, as if none of it existed. Indeed he was beginning to look monumentally out of place until, with more of a spring in his step, he led the way upstairs to a cool saloon full of crisply laundered tables on which silver glittered and glass shone. Here, harassed and hot, he sat down to a large menu with larger print and the town's largest prices, ordered a pastis and visibly relaxed when the drink came at once. I often noted his infectious power of visibly relaxing; we, his companions, felt better too. Quite quickly he ordered rather too many courses, all far too rich, with wines to match. ('Now to begin could we not drink a bottle or so of that white Châteauneuf-du-Pape, and to follow . . .') Beginning to exude high spirits, he turned his thoughts elsewhere again.

He was five, still at Norwood. 'One day at that appalling school there was a snowfall,' he said, 'and with a huge magnifying glass they showed us snowflakes on the window. And suddenly all that grey slush dropping against the cold miserable London outside was transformed, beyond anything I had ever seen or imagined, into something beautiful. It came back to me as one of my very few memories of that time when years later I was reading Marcus Aurelius on the nature of beauty, when he attributed that indescribable quality even to the foam on the jaws of a wild boar, and at once I saw those snowflakes again outside the school window.' He paused. 'Inside, however,' he said with a slight leer, 'sex was rearing its even more beautiful head. I had a passionate love affair with a Swedish girl, a little blonde. We called each other husband and wife – think, I was actually imitating at the age of five what for most men remains a lifelong fantasy. Our parting was certainly worthy of any Swedish film epic. We kissed each other and swore eternal devotion and then I caught the train to Bournemouth, that asylum of the broken heart.'

The first course came. Served off salvers, and closely watched in every detail by Durrell, it consisted of thin slices of salmon in a blend of cream and white Châteauneuf, the wine we were now drinking. There was a second's pause, the length of a sigh, as if for a silent grace. Then he fell to, but with enormous care, ramrod straight, on the edge of his seat, putting down his knife and fork between each mouthful, an expression of glazed reverence in his eyes.

'Mmmm,' Jacquie said. 'It's good, isn't it, Gerry?'

Durrell nodded gravely.

Not another word was spoken. Glasses were lifted with cordial solemnity. The food was delectable. It was as if the entire morning had been a preparation for this sacrament. Painstakingly, Durrell wiped up the remainder of the sauce with a fragment of bread. 'Did you enjoy that, Gerry?' Jacquie said in slightly stronger Mancunian than usual. He nodded, satisfied. And began talking again.

The train rescuing the family from London had now arrived at Bournemouth. Their move to a virtual palace – 'the type of heavy mansion you now see rotting gently away in Deauville' – was the result of Mother's usual brand of logic. She had to have room to cope with the rush when the children brought hordes of friends to stay for weeks. For the bulk of the year these extensive apartments, staircases that led nowhere, warrens of attics, weird cellars full of cobwebs, all set in a garden ('with room for fifty modern bungalows') shrieking with songbirds, were occupied only by Mother, the housekeeper and Gerry. The others merely dashed through it from time to time. It was their corridor to the next event. Literally a folly, the house was yet another nail in the coffin of the family finances.

The move increased the boy's isolation and confirmed his shyness. It also enlarged his horizons. The garden – orchard, pine copse, a sunken lawn 'big enough for four tennis courts' – was in effect a private nature reserve, a first laboratory for someone still learning that animals were proprietors of the world, outnumbering us by thousands of species to one. Being here with all this space intensified that decision already made at the age of six in London, to have his own zoo.

He muttered this ambition to Mother as they took their occasional walk on the promenade. 'Yes, dear,' she said in a patient tone at intervals. Meanwhile he described exactly which animals were to be displayed, how they were to be caged, and most of all the accommodation which the cottage, where she and he would live, must offer. 'I was obsessed with this cottage,' Durrell said, 'and I've been obsessed ever since. The nearest I approached the ideal is the manor house in Jersey. But though I was intelligent enough to realise that you couldn't have a zoo without a place to live yourself, I never thought in terms of staff. I had every intention of feeding and cleaning out five thousand animals a day, all by myself. I must have been something of an idealist.'

There was a further pause to make room for another course. This time Gerry's reflective ritual was applied to a duck, a wild duck shot in the Camargue, a plump duck swimming in rich brown sauce. While his brain brooded over what was happening to his palate, his eyes focused on the legs of distant waitresses or bourgeois families equally intent on their Sunday lunch. Minute quantities of precisely cooked vegetables accompanied the duck. For an appreciable time after the carcase had been removed the talk was of how exceptionally delicious the duck had been, in what ways, and whether it could ever be bettered or even rivalled.

'Why don't we have another bottle of this excellent stuff?' Durrell said.

In Bournemouth there was a school called the Birches at the end of the road, run by someone known as Squig. Like all good teachers, Miss Squire taught what the boy longed to know, introducing sticklebacks into the aquarium, spicing her instruction with animal lore as soon as she guessed her new pupil's obsession. 'Every morning I used to trot down with a matchbox of decaying slugs or woodlice,' he said, 'and she'd greet them with cries of joy, how beautiful they were, and didn't I think they'd be happier in the flowerbed until break – and then the entire break would be spent trying to recapture them. But she accomplished it with such charm that I never realised what the trick was. You see, the need to share my pursuits didn't occur to me – I just assumed that everyone was interested in animals. I thought the entire world joined in my deep and overwhelming desire to get on intimate terms with a snail. I still do. You see how selfish I was. I still am.'

We strolled back to the square in the afternoon heat. The jeans-wearing youths were still much in evidence. 'Man is an unattractive mammal,' Durrell murmured testily. In an open lorry a band of trombonists and drummers, attired as nuns and clowns, drove at a funereal pace past the war memorial, advertising a function at Villeneuve called theatricide. Durrell sat down outside a café, ordered a beer and used yesterday's *Daily Telegraph*, which he had no intention of reading, to shield his eyes and conceal his distaste. 'These are the very people for whom I'm slaving out my guts,' muttered the voice of the conservationist.

We drove back to the *mazet* through the meridional lull, stupefied by refreshment, aching for siestas. Durrell was selfish; whether or not hoping to be contradicted, he had said it himself. He also reported that in the view of others during those Bournemouth years he had displayed no signs of the spoilt brat. On the surface there was nothing precocious about him because he took it all as his due. He still did; nobody seemed to mind his lordly airs. The intensity of his interest in a single vast issue was in his favour. His vision of the world's future was entirely at his own expense, inconveniencing him at every turn, wrecking his bank balance, disturbing his peace of mind. The small boy listening to a thrush in the garden while he picked up yet another snail for the matchbox had been already in the process of turning himself, to everyone's awe, into a one-man society.

Shortly after their arrival in the Bournemouth house, 'where masses of little staircases led to one room', Mother was swept off to a nursing home with a nervous breakdown, no doubt caused by her husband's death, leaving her with too many children and too little money. Against Larry's advice a housekeeper, 'hand-picked by some of our awful relatives', was put in charge. Her nightly habit, since she loathed being disturbed after retiring, was to lock Gerry in his room until morning. The result was that the boy awoke with bladder bursting, held it in, dropped off to sleep and wet the bed, a misdemeanour never committed in his earlier childhood, but to which he attributed nowadays the phenomenal restraint of his bladder. For that morning, as sharp reprimand, he was forced to swallow a soup in which semolina floated. These weeks away from Mother were the only shadows in his first ten years.

Back at the *mazet* Durrell stumbled off to his bedroom. Jacquie lay down under a rosebush out of the sun and closed her eyes. Alone in my room with a mosquito, I sorted into chronological order the Durrell paperbacks I had brought with me, resolved to read a few paragraphs, then dozed off. It might have been more to the point to take a look at *The Wind in the Willows*, Gerry's first brush with the printed word. It had been read to him nightly by Mother in London, over and over again. When his appetite for those animal adventures in an ideal world showed no signs of diminishing, she pointed out that a bit of effort on his part would mean he need no longer depend on her for the magic. Learn to read, she told him. Take the book into your own room. Spend day and night with it. Live alone in someone else's world. 'I can remember being slightly confused,' he had said earlier that day. 'I recall looking at letters and wondering why, when they all joined together, they made a sound. I couldn't grasp the fact that little marks on paper in a row were in fact noises that made sense. That was the hard part. Once the relationship clicked, of course, it was simple . . .'

Another factor in his fast response to language was a habit of Larry's in London. He used to come back in the small hours from a jazz session where he played the piano and sit on Mother's bed, smoking a cigarette and chatting. She could never sleep until he was in. Gerry's bed was in the same room. He invariably sprang awake, avid for Larry's boozy talk, the wit that twisted events into an epic or a joke, and then had to have a story read to him, or invented on the spot, to put him back to sleep.

On one such occasion Larry had just composed the tale of a man who went mad in a room with cherries on the wallpaper and disembowelled himself with a razorblade. Larry lay on the bed reading it aloud. Mother thought it not quite suitable at that time of night, but to the boy it was vivid, sympathetic, in no way horrifying. He valued Larry's almost casual way of plunging him in at the deep end of education: holding nothing back, never censoring, showing the world as it was. 'On the rare occasions when I've had to do with children,' Durrell said, 'I've tried to use the same technique. And you can see their minds opening.' Larry, for whom in the care of his fatherless brother nothing was too much trouble, made it his responsibility to suggest other horizons, stir his imagination. A bed of roses was never only a bed of roses. It was a parade of virgins waiting to be raped or a ballet troupe swaying in the breeze of applause or faces in a crowd puckered with anxiety.

So Durrell learned to read in Bournemouth at no pre-cocious age, but with a sense of intense discovery. The classics of childhood dropped into his lap as ripely as plums, Kenneth Grahame, Edward Lear, Lewis Carroll, all making animals human. Girls lived in clouds. Castles made of glass materialised out of mist. Gerry tucked into the fairy-tale images that always appeared to mean more than they said. He neither wanted nor cared much for other children. To this day he had no particular need for people, feeling a timidity in their presence, a slight fear. But the books nourished him. So did the huge garden. All he dreaded was the regular visits of two 'domineering hags of cousins' who goaded, teased and sneered. Day after day they compared his toys with theirs, measuring their superiority by the variety of their possessions, until after hours of silent suffering the boy said, 'You haven't got a father in heaven, but I have', which satisfied him as repartee but failed to quell the cousins.

When I got up, the Durrells were already busy in the kitchen. The next day, after a farewell lunch at the *mazet* for various local friends and expats, they were leaving for Jersey. Summer was on the way out, another paradise ending. Drugged from his siesta, Durrell was making an effort to confront his menu for the morrow. Disposed on the table, as on an altar, were supermarket packages of raisins, pine nuts, spices, baby pots of herbs. He took his cooking seriously. The kitchen was apt to become a private temple from which others at these moments must be banished, including Jacquie, who was good on grilled chops, eggs and bacon, roasts, the daily stand-bys that gave Gerry all the nourishment he needed to be inventive.

Durrell preferred the exotic. For the balance of tastes in a single dish, a stronger flavoured (or masculine) approach was needed – in silence, without interference, but with love, even if it ended in a chaos of unwashed pans, spilt ingredients and Gerry's conviction that somebody else ought to clean up. But the result was likely to be good. 'When you spend a lot of your time, as I do, roughing it,' Durrell said, smearing quantities of butter over the breasts of a turkey, 'you want home to be utterly civilised. Mind you, I try to be civilised on collecting trips, but the definition of luxury in the jungle is whether you have a deck chair or not. In France civilisation goes without saying. The good things in life are the blood in the veins, the heart of the matter.'

His brow furrowed; he had forgotten to switch on the oven. He waved a hand airily. The turkey could wait. The pause was a perfect excuse to drink a glass of chilled wine. 'I adore France,' he said. 'On the other hand I love the Greek *joie de vivre* – a French phrase, why haven't we got one like it? A Greek lives every moment of his life as intensely as a dog in a forest of lampposts. In a French village they ask you things politely about yourself, but respect your privacy. Greeks will demand within five minutes of knowing you how often you have sexual intercourse with your wife. And they'd tell you about their habits if you asked. Being British, you never do.'

He gazed into the kitchen grate where a log burned. When he was eight, quite suddenly, the family upped sticks in Bournemouth and went to live in Greece. With a sigh he picked up the roasting pan and thrust the bird into the oven. 'If I ever get round to writing my cookbook,' he said, 'it will be called *Cholesterol Cooking*, every recipe guaranteed to give you a thrombosis. Yes, I feel like a party. Again it's purely selfish, because it gives me enormous pleasure to give pleasure. It's such fun buying the cheese and getting it to the right stage, so it smells like the socks of the entire Boer War advancing on you.' Inside the oven the turkey started sizzling; outside the sun began sinking. Durrell stood for a moment in the doorway, bestowing a final glance on the valley, which would soon be dark. 'There's a threat to living standards,' he said, 'in the yaps and screams of socialism, which is basically all right except that it doesn't work. Instead of curing problems of world population, socialism concentrates on bringing everyone down to the same standard, instead of helping everyone to rise to a higher one. You can't find anywhere to be alone. You feel it's wicked to have a bottle of Scotch if there are some who haven't, so in the end no one has anything at all.'

He turned indoors. 'I don't want to be absolutely equal. I want to be me. If I choose to spend a hundred pounds on a party, I'll do it. I've earned it. Human beings can help themselves, they'll get no help from me. If I live to be seventy, the whole deteriorating mess will just about see me out. There has been, and will continue to be, radical change, but with nothing to counterbalance it, no sense, no creation, no wisdom. The Third World! I ask you! They invent a phrase which divides us all just when we're trying to get together, a journalistic phrase like the Abominable Snowman, implying something other than it is, trumping up something that doesn't exist. As far as I'm concerned there's only one world, and I'm in it and it's inside me, like the duck we had for lunch, and those louts in the square, and Jacquie, and the handful of mountain gorillas getting fewer every day, and wine, and my horrible cousins, and MPs and Mexicans and Chinks, and the memory of my mother, and Norwood, Bourne-mouth and India, and Larry writing his incomprehensible masterpieces, and breeding white-eared pheasants and anything human, animal and alive.'

These last words were snapped. He turned back to his cooking in a self-induced rage, which grizzled away into silence. The outburst might have sprung from forgetting to switch on the oven, or a headache, or not being alone in the kitchen. With him you never quite knew whether you personally were to blame for such sudden squalls, even when the principle he was defending was as ostensibly watertight as today's. When these moments caught him in the gut, a lot of bile could spill out, rather as though it had been saved up to put an offender in his place. So this was a good moment to do a bit of homework. I had Durrell to read or transcribe in the privacy of the guest room, nicely cut off from the main body of the *mazet* up a flight of tiled stairs at the back, an austerely furnished apartment ideal for 'my biographer', the title Gerry was now dryly giving me on introduction to strangers.

I had read most of Durrell's books in the past, usually when they appeared. Now at night — in France the household tended to go early to bed — I thought I'd look at them again. I wanted to know what the books truly, perhaps cruelly, said about their author, the odd hint or touch of the sinister one might miss when reading them solely for the pleasure of his company. Then, because a book no more existed without a reader's reaction than a play in an empty theatre, I felt a need to be precise about why these books appealed to me, what the secret was, if any, and what in turn they told me about myself. So this was not to be literary criticism. I also wanted to avoid repeating in my pages material, however juicy, out of which Durrell had squeezed the last pip in books of his own.

I put out the complete works to date, measuring more than a yard, on a long shelf beside the bed, lay back to consider them afresh by staying awake into the small hours, and fell gratefully asleep. It must have been a second hangover from that lunch. I had no recollection of anyone calling me down for supper.

Monday

THE DAY of the farewell party dawned. Just after eight a blue Volkswagen dormobile, known as the Escargot, drew up at the gates and hooted. A stocky figure took a bus driver's jump down. Lawrence Durrell stretched, uttered a casual greeting, eyed his bleary brother, nodded in my direction, and at once began gassing. He mentioned Bergson, which drew him into Proust, which led to a sketch map of the Normandy coast, followed by an attack on English manners, and who on earth was coming to this party?

Nobody gave much ear to Larry. This was the usual flow of ideas, delivered with the petulance of a steely intellectual, expecting to be challenged. No one contradicted him. It was too early. Brotherly love glared at him a bit, then Gerry said he was going inside for a shower. With a sharp look at me, wondering if philosophy was worth pursuing, Larry sat down, tucking his legs under him in the half-lotus on the bench by the table. His hands were trembling a little as he gestured, lips compressed: had Proust (or perhaps Bergson) ever come to Sommières where Larry now lived? It was too early in the morning for such pure speculation. 'I must be off,' he said, not moving. Gerry returned, dripping wet and dressed in his usual morning towel, and realised in an instant that his brother wanted to be offered a drink.

Larry had been up since five, first practising his yoga in the huge shuttered mansion alone. Three hours at his papers thereafter, completing a day's work by anyone else's breakfast-time, then the bottle: his way of protecting his sanity, he claimed, against assaults of gloom.

'Try this white wine,' Gerry said, brandishing some local stuff we had sampled the day before and not much liked. 'If it's not to your taste, there's some gin and things.' He spoke in a restrained manner, with no trace of his usual gaiety. The presence of his elder brother seemed to awe or irritate him, as it might a child. At fifty he looked pudgier, older even, than Larry thirteen years his senior, and clumsy, compared with the taut miniature figure of the novelist. Their eyes suggested a quickening interest in each other, as though they had just been introduced.

It was another unsurpassable day, the sky a silky blue, the sun dazzling. It was in pursuit of such mornings that in 1933 the Durrell family, at Larry's instigation, quitted arctic Bournemouth and went south to Corfu with a lot of luggage and insects. From this timeless paradise (a phrase never quite avoided by either brother) only the coming war in 1939 drove them away. Now, thirty-five years later, they sat together in similar sunshine, each talking, neither quite listening. Within minutes two glasses of wine had slipped without comment past Larry's busy larynx. A mood of all the chances missed caught them, as though mention of Corfu, long after ejection from that Eden, were too painful for either. They went quite quiet for a while. I wondered if this melancholy was of the cast well known to writers, even well-loved by them, indeed essential to their function.

Gerry heaved a sigh and padded behind the house for a swim. A heavy splash was heard. 'In Corfu, do you see,' Larry said, 'we reconstituted the Indian period which we all missed. The island exploded into another open-air time of our lives, because one lived virtually naked in the sun. Without Corfu I don't think Gerry would have managed to drag himself together and do all he has achieved, and of course he celebrated it in *My Family.*

'In all his work there's nothing quite as lyrical, quite as well-formed, as perfect – I was going to say it's his Dante's *Inferno*, but I meant the opposite. I reckon I too got born in Corfu, in the sense that my first poems and publication of *The Black Book* date from that period. *Prospero's Cell* was a posthumous attempt to tip my hat to the place, but it was really that spell between the wars that was – you can only say paradise.'

Larry hardly moved, stared into space, sipped his wine. 'He's remarkably tenacious,' he went on. 'You see now this Herculean chunk of a hero with a beard. But Gerry was enormously fragile as an adolescent and physically timid. Frail, slender, he resembled my father, who was tall too. But the rest of us were all tiny, chunky, like coal miners, with that kind of resilience that comes only from being troglodytes. Gerry was always lanky, his shoulder-span small; he wasn't a boxer and he couldn't curl up like a toad and take the weight of the punches. Also psychically he was nervous, jumpy, shy – and all this he's managed to convert into a powerful pile-driving character without losing the sensitive side, which is marvellous. We've all had strains and stresses, but Gerry has had a much harder time of it than any of us. It's a quite remarkable feat to turn yourself into a toughie while remaining friable and tender inside.'

The two brothers had much in common. Their faces shared a family look of spry mischief. In both a steady glitter of humour in the eye was always on the verge of mockery. Their laughter, at only slightly different pitches, was an infectious staccato. Not only did it punctuate the anecdotal style they both favoured, it ensured its success. Gerry deployed this storytelling technique in his books; Larry held it in reserve for social use.

Serious men of like temperament, they were different only in emphasis. Literary to the bone, Larry was a nonstop talker, life having value only if converted into the scintillating currency of words. Festooning life with phrases made it tolerable. Nothing so intellectual inhibited his brother's thought. Ideas never spilled out of Gerry, tumbling over one another in a quest for patterns. They emerged from his intuition at due intervals with a force not to be brooked. Larry debated with vivacity, never crediting his own convictions enough to be hurt by attacks on them. Gerry, if opposed, was more likely to retreat into speechless anger. With him you could never question his version of the truth without arousing his scorn.

They were always twitting each other about their books. The only thing that weakened Larry's work – highly wrought poems, novels that in the Sixties fed a Western world starved of colour and myth – was the absence of faith, even in Western values. Larry created a world of his own, only to prove it hollow, letting you in on the luxury of not taking it seriously. But in his brother's easier books, indeed in Gerry's every activity – whether breeding gorillas, smelling a sprig of thyme before sprinkling it into a soup, gazing intently at an ant – you always found conviction. He might take a teasing attitude, for fear of being thought stuffy or pedagogic. But a faith was always there: in some goodness in nature, whatever its evils, a pleasure in the life force, despite the trouble it caused.

As the morning sun penetrated the bamboo, Larry looked up thoughtfully into the veranda's tangle of passion flowers. 'You see, he was softened up by his ma,' he said. 'While we were all catapulted out into the world and had to get over our mother fixation somehow by force majeure, he was kept at home. He never had to fight with his fists for room to breathe, as we had to in various schools and miles from home. Gerry tagged along with ma, which was very weakening, but also very enriching. When it came to putting on his own show, making his own world work, he had to struggle against the enormous indulgence granted him in living with this extraordinary mother, the most charming creature you could imagine, the most demanding in affection. But Gerry has overcome this weird childhood and turned himself into a man, and a tough one, and a fully developed one, in the teeth of an upbringing that might well have justified no career at all. It's a triumph.'

Larry finished the bottle of wine, announced his intention of returning for the lunch party, and stalked off to the dormobile. I accompanied him down the steps. 'As a small boy he was impossible,' he said. 'A terrible nuisance. He has recounted the worst of himself as well as the best in that *Family* book. Oh, it was matchboxes full of scorpions all the time, I didn't dare to sit down anywhere in the house, and of course Mother was there to defend him – the slightest criticism and she would snarl like a bear, and meanwhile there were beetles in the soup. No,' Larry said, hauling himself into the driving seat, 'he was intolerable, he needed to be thrashed.'

On the terrace, as I rejoined him, Gerry was already expostulating over Larry's absurdity. The poor fellow was obsessed with money. His private life was byzantine. For a moment Gerry wondered aloud whether the moment had arrived to 'press something to the left kidney', but his gaze drifted off into examining the quality of the light. This valley in France seemed heavy with ripe fruit. It might have been a Corfiot morning long ago.

In Bournemouth Gerry had kept asking Mother about Corfu. She, browbeaten by Larry, who was already abroad and had taken command of the operation from afar, knew as little as the boy did. But why go? Why uproot? Why not stay? Why travel? He nagged on. For some reason he saw Corfu as an oval rock, long and grey and bare, with a few houses perched on it and no trees, 'a tropical Iceland without the volcanoes'. He insisted on including in the luggage his boxes of stones and shells, his precious eggs — 'rather like a dying man making his last request' — because he feared there would be nothing to do on the island. These treasures of his might help to keep the family amused. Later, when he knew better, Gerry found that the Greek peasant children had a similar vision of England, grey and wet and drab. Their ignorance was more accurate than his.

Thus, caring more for his egg collection than for the white cliffs, Durrell sailed in a Japanese ship through the Bay of Biscay, touching land only at Gibraltar where, thanks to Mother's determination to pay a social call on an elderly friend on the rock, they almost missed the boat to Naples.

The weather turned breathlessly blue, the sea sparkled, but the much-travelled twelve-year-old was not impressed. He dimly recalled such conditions on the trip home from India as an infant. So he kept his head down. Not until they reached Naples was he forced out of his preoccupation with pets and collections by the noise and reek of back alleys teeming with children, meat, trinkets and a hunchback who spoke English. 'Anyone who spoke English abroad,' Durrell said, 'was leaped upon by Mother with cries of joy, even if he had one leg and a parrot on his shoulder. Should anyone suggest the hunchback might be cheating her, she'd say, "Yes, dear, but I understand what he's saying so don't make a fuss." That summarises her attitude to life.' After a swift and costly tour of the sights of Naples ('we nearly did die'), the hunchback escorted them to a compartment with horsehair seats, where they fell asleep as the train jogged towards Brindisi. When in the small hours a portly Italian joining the train was nuzzled in the stomach by Roger the dog, the man thought it was Mother and started gabbling like an operatic seducer.

At dawn in Brindisi, a man with 'Thomas Cook' printed on his cap and a minimum of English, materialised like a genie out of the confusion, at once understanding Mother's need for a hot cup of tea. He bowed them into a hotel of tottering charm, all potted palms and chipped statuary, where they spent the day draining the dregs of an England gone for good. They caught the night boat to Corfu.

First light revealed a shape on the horizon not unlike Durrell's bare grey fantasy. Then, as the sun blazed into the early sky, he stood at the rail, 'intrigued by the caves and beaches passing the ship, excited, but still a bit frightened, confused . . .' On landing, and for the next month, a muddled, irritable, penniless boredom reigned in the family. The bank in London had failed to send money. They were dependent on the charity of the Swiss lady who ran their hotel, although, Durrell said, any Greek would have done no less. 'The English were revered in Greece at that time, as gods almost, though God knows why.'

Openly homesick, sister Margaret wept. Gerry wailed in sympathy. In the dim saloons of the hotel they all sat squabbling about what to do next. Here was the first moment of insecurity since their father's death. They had to borrow cash for the motor-car required to inspect any villas on offer. All Durrell noticed in that month was that his black-and-white world had been converted, reminding him of the midway change in *The Wizard of Oz*, into glorious technicolour. Even wandering down to gaze at the shallow water lapping at the base of the fort introduced him to a sea shimmeringly different from the grubby waves of Bournemouth. So there was hope. And then there was Spiro bursting on the scene, Spiro, the saturnine taxi driver, one of several lovingly portrayed heroes in *My Family*, who at once found them a house.

'It wasn't until we moved into that first villa,' Durrell said, 'that suddenly we realised we had been transported into paradise. Think, we had been standing at the gates bitterly bemoaning the fact that we were there! And then, oh, the first day of moving in, clearing the mountains of belongings from the customs shed, Mother having packed all the sheets and frying pans and jugs, all the excitement of having two car loads carried by peasants screeching an incomprehensible language up through the olive groves, opening of shutters, sweeping of floors, making of beds, everyone telling me to get out of the way because I was raising dust or asking questions. Eventually I trotted off quietly to the tiny garden with a big privet hedge, neat as neat, as tidy as a Victorian moral, stones marking out flowerbeds patterned as stars, circles, squares, the interlocking gravel paths richly overgrown, stuffed with spring flowers and consequently spring insects, creatures I had never seen or imagined before. This was a tremendous moment of revelation for me – I'd never thought of such fecundity: this garden overloaded with plants, every stone I turned over had twenty different creatures under it, and there were huge blue furry bumblebees flying round my head and praying mantises staring at me even more astonished than I was, and for me it was like being pushed off the Bournemouth cliffs into heaven. From then onwards, just like that, I was home.'

His recollection of Greece was always intense. In talking of Corfu he seemed not to indulge in nostalgia but to relive in total recall those moments as a youngster. 'Other countries,' Larry wrote in *Prospero's Cell*, 'may offer you discoveries in manners or lore or landscape; Greece offers you something harder – the discovery of yourself.' As Gerry brooded on this aperçu, a knife-edged Mancunian voice from the kitchen reminded Durrell of the difficulties of preparing lunch single-handed for ten or so people. He, on the other hand, was supposed to be skilled at the art. With an irritable intake of breath and a brief heavenward lift of the eyeballs, Durrell plodded dutifully indoors. A hint of altercation was at once wiped out by loud hammering. He was probably liberating a lobster from its shell or crushing garlic.

From the terrace I stared out at the landscape which in talk Durrell had already invested with something of the dazzling atmosphere of his childhood Corfu. What was really the essence, as far as it could be recaptured, of that island? Nobody knew better than a trim, gentle figure, nowadays living in a London suburb where I had talked to him a month ago, who had a short white beard and exactly the precise speech littered with hesitations, the scholarly modesty, the speculative concern for the living world, that Durrell attributed to him in *My Family.* He was drawn with affection as a character of touching comedy, almost a magician. In Dr Theodore Stephanides, then in his forties, the boy had found the perfect mentor, not only an authority on the island and a Greek, but a mind in which poetic response had achieved a nice balance with scientific research: a blend of the two cultures which Durrell had ever since been striving, sometimes with impotent fury, often with success, to reconcile in himself.

Theodore had first come to Corfu in 1907, when that paradise was much as Edward Lear portrayed it in his dainty sketches. There was one motorcar. At Paleokastritsa, a resort now infected by rashes of hotel and villa, there was nothing but a monastery and a shack on the beach that sold lobsters. Forced out of this quietude by the First World War, Theodore spent two years in Marseille, served with the artillery under Venizelos in Macedonia, and, only after the Asia Minor war with Turkey, proceeded to Paris in 1923 to begin his medical studies. On his return to Corfu he started his own practice, later to become the only X-ray facility on the island. Meanwhile he was drawn into research for a work on the freshwater biology of Corfu. For 'the arcane professor of broken bones', as Larry called him, the pattern was set. By the time the Durrells appeared on the scene, this timid expert with his quirky humour, his almost English irony and outrageous taste in puns (murmuring, when consulted about the diet of seabirds, 'all the nice gulls love a sailor'), was just the person for Gerry at just the right time.

'You could *feel* it soaking into you': Theodore's words to describe the potent effect of the island. It was a place where, according to Larry, time was no more than a word and distance was measured by the length of a smoke – 'Ask a peasant how far away a village is and he will reply, nine times out of ten, that it's a matter of so many cigarettes.'

I remembered that Theodore had told me a number of

stories to convey the nature of the island's spell. There was the case, for instance, of a large wooden barrel which had been adapted as a rostrum for some dignitaries at a public function. Typically no one had thought to calculate their combined weight. To the amusement of all, the makeshift platform caved in – but without causing a single injury, except to *amour-propre*. On another occasion the local Boy Scouts were demonstrating how to destroy a pontoon they had built. When the crowds closed in to watch the display, an immense explosion shattered every window in sight. Someone had substituted a live dynamite cartridge for the blank. Again nobody was hurt, and no outrage was expressed by the audience, beyond suggesting that someone should have warned them to stand further back. Theodore also mentioned the insane moment when two charabancs decided to race each other down the narrow hairpin bends on the road to Paleokastritsa. The best coach won. Its rival plunged into a ravine, which cost one passenger a broken leg while another almost drowned in the only puddle that had survived a particularly dry summer. Theodore's anecdotes proved the point: why bother to exaggerate Corfu, when the island exaggerated itself?

Engaged as an informal tutor, Theodore was at once impressed by the boy Durrell's enthusiasm and energy. Gerry had a capacity to invent new ways of collecting specimens or preserving them, to deduce by intuition, to home in on observations that had eluded his teachers. Theodore found that the essential qualities of a naturalist, equally those in his view of a well-rounded man, were present in Durrell from the moment he first set foot on the island. The prime quality was tremendous patience. The boy lost all sense of time when on the track of something. Perched utterly still in the branches of a tree, he stared for hours at the life of some unhurried creature. And he took animals on their own terms, the hardest discipline of all for anyone in serious quest of knowledge. No icy science intruded on Gerry's approach to his quarry. He never wanted to possess the creature by pinning it through the thorax or shooting it dead. He preferred to identify with the way it lived, both as an individual and in relation to others. The thought never occurred to the boy that the death of something would tell him anything.

At this age, said Theodore, Gerry also had immeasurable curiosity. Unconcerned for his own comfort, in a pure spirit of enquiry, he never gave up until he knew, provisionally, the answer to the question – about an owl, a rat or an amoeba – which hammered at the front of his mind. Theodore felt that for Gerry this curiosity was its own reward, while it also led to further understanding of life's basic patterns: if the amoeba had failed to reproduce, to put it extremely, none of us would be here to observe it. Another fact Gerry recognised early was that all species, and every individual within them, had an equal right to exist. Not only could the boy Durrell walk Theodore off his feet, but his dedication was amazing in one so young, never missing a thing, however small, never giving up. Also astonishing, thought Theodore, was the way animals responded to Gerry. All creatures great and small, even as small as a rat, took to him as if sensing they were safer with predators removed and diet secured than in their natural state. They admitted his presence either excitably, as they would a mate, or quietly, acknowledging a friend. All this struck people as improbable, until they were lucky enough to catch Durrell in the act of dealing with animals.

So, on still afternoons deafened by cicadas, the two of them set off into the olive groves that descended towards the sea: 'sallying forth', as Theodore phrased it, after a light lunch, their goal a pond or lagoon not far from the villa. Each was armed with a dipping net of their own design, a cone of fine muslin ending in a small glass tube at the point of a long stick. Swept through the water, the muslin trapped the tiny creatures, which then slithered into the tube, to be identified with a magnifying glass and recorded in a notebook. Over their shoulders hung knapsacks filled with the boxes, bottles and bags to take specimens home for further observation. A few large canvas bags were included for bulkier items like pond tortoises or water snakes and smaller containers packed with damp moss for frogs and lizards. These also served to return previous captures to their habitats after sufficient study. On rare occasions they brought a butterfly net. 'But both Gerald and I,' Theodore told me, 'were more interested in studying *live* creatures and kept our collection of preserved specimens to a minimum. We always carried with us a bottle of fresh lemonade and a supply of biscuits or sandwiches for the inner man.'

The scene then shifted surreptitiously to the villa, where additions to the menagerie could be sure of a cool welcome. Proceedings began, Theodore said, with a commando raid on Mrs Durrell's kitchen, where they requisitioned every soup plate in sight, 'often making ourselves somewhat unpopular with that good lady'. Then the specimens were sorted out with a battery of teaspoons (also 'borrowed') and glass pipettes (their own). At this point, microscopes in the vanguard of the attack, the creatures were accommodated in 'a whole army corps of aquaria' from jam jars to the plump bottles in sweet shops. All these containers, a layer of fine gravel at the bottom, were aerated by sprays of pondweed. Once again the battle for knowledge – Theodore kept resorting to military metaphor – was launched in earnest.

I reflected peaceably on these warlike details, as something of the magic of Corfu made itself felt on that restful terrace in France. It seemed neither a holiday island nor a dream nor a retreat from the world, hardly even a paradise. On the contrary, it possessed and retained an actuality that nothing could destroy in Durrell. After two books that relentlessly plundered his memories of the island, he was still in thrall. He behaved as if nothing that had happened to him since measured up in immediacy to Corfu's charm. I often detected an undertone of despondency in Durrell's mood. It may have sprung from the wonder of that spoilt boyhood which was bound to end in tears.

I recalled also Theodore telling me that, with an innocence that seemed odd in these later times, the boy had been taught two principles: that life, left to itself without human interference, maintained its own balances; and that the human role in the scheme of things was one of humble omniscience. Theodore epitomised that humility. He viewed nature as a source of wonder. But behind his modesty lay the supremacy of the scientist – a scientist, at least, up to 1945 – who by example allowed a small boy to feel that the known and knowable world was within his power. Thus the happiest days of Durrell's life had been bound to lead to sadder ones. At a tender age he saw a complete world. He had spent the rest of his life discovering gradually how incomplete it was, how flawed, how close to ruin.

No wonder Durrell was now clinging to food, drink, sunshine and peace, in an effort to convince himself that all was partly well with the world. He did not believe any such thing. Nor did he believe that any effort on his part to be hospitable to the almost extinct helped life much. 'My achievement,' he once said with a sneer, 'has been like chipping away at the base of Mount Everest with a teaspoon.'

In the kitchen Durrell was chopping onions, his whole body bent to the task. 'I like cooking for people who adore food,' he said. 'They allow me to give myself pleasure, whereas most friends are too busy trying to get pleasure out of you. But just to say, We'll give a party! It's exciting, what a moment – a party! And then we sit down and work out who's not speaking to whom, and whether so-and-so is still having a ding-dong with somebody's wife, and then we settle down to the menu. We'll start with this and that, but dear old what's-it is coming and he can't eat it – rather like Theodore, who comes out in nettle rash at the mere mention of shellfish, and to whom I once by mistake gave a prawn curry so subtly disguised that he not only pronounced it delicious but retained his health. I love producing for starters some esoteric dish – on my birthday once I sent to Norway for smoked reindeer, served with melted butter, horseradish enriched with cream, rolled up and spiked . . .'

Durrell paused to tip his chopped onion into a salad bowl. 'Then soups; I consider soups a much neglected gastronomic area. People think you merely hurl a couple of cabbages into a feeble stock, but just think of the aristocratic glory of a *bisque de homard* or a fish soup the colour of an African maiden. I sometimes run up a most exquisite *potage* of lettuce, watercress and so on, which in summer I ice, but I suppose it would never be popular in England because they never have any summer.'

Knife in one hand, glass of wine in the other, Durrell

stood tensely, as if summoning up a vision. 'Then, a last look at the table,' he said. 'Everything aglitter, the white froth of napkins, glasses in place, a civilised polish to the whole, the juices already churning away as though the kraken were awaking in the lower gut. You open the wine, corks plop, aromas rise, and there's the splendid palaver about what to drink with what, and whether that rotund, full-bellied, head-hammering red you bought at a little cave in the Côtes-du-Rhône will go with the game, the quails stuffed with raisins and pine nuts, the hare (which I adore, I've invented a method with red wine and olives), the chicken with groundnut chop. At the drop of a grouse I give people game and I've got a lovely venison recipe from 1525 to which I've added inimitable touches of my own, but you need at least fifteen people to eat it, and they take days to recover. You steep the venison, you see, for a good forty-eight hours before you bring it anywhere near the heat and it's so fragrant it almost walks off the bone and embraces you. Cooking *is* sexy – the way to a girl's bedroom is through her stomach.'

Reaching this climax, Durrell relaxed and put down the knife but not the glass. 'My reason for liking parties,' he said, 'is that they were so marvellous in Corfu. We used to start lunch at twelve noon and end at three in the morning, eating the cold remains on the beach after a midnight bathe. It was my mother's training. As well as being the most generous of persons, she suffered from a flamboyant streak which always convinced her that one chicken wouldn't be enough, so she roasted three just in case. In Britain you can no longer afford to give joy to your friends – or to yourself, for that matter. A party nowadays is measuring out a drink to a couple of people and marking the bottle of Scotch with a pencil. There's no scale. You really have to give a banquet, not a cut off a cold joint.'

Durrell put the finishing touches to his salad and walked out into the breathless noon heat of the terrace. No doubt he saw this scene at the *mazet* as a present echo of a villa in Corfu, the promise of lunch at the long table, the bottles open on the ice. The sky was as gentian blue as he had described it, the smooth enamelled blue of a jay's eye, the very blue of the tiny flames that devoured the olive logs in the charcoal pits: the books were rich with phrases describing just such skies as were above us now. Out of pearl-white sands, on just such a festive day, the trumpets of white lilies were growing, as seven fat snipe were roasting on an olive-wood spit and 'prodigious quantities of wine', as drunk by Greeks on all occasions, had been put out to cool in the sea. 'The magic settled over us as gently and clingingly as pollen'; the family lay about on beach or veranda, 'eating, reading, sleeping or just simply arguing'; Margaret stumbling into the comfort of her novel proverbs ('a change is as good as a feast'); Larry constantly asserting his comic superiority, falling into swamps, drunkenly taking charge when the house was on fire, lolling like a pasha cushioned by cultural references; Leslie oiling his guns, losing his temper, shooting anything that moved. And always there were parties, springing out of the dawn and dying away in the dusk, balloons floating over Theodore as he launched into a Greek dance or told stories nobody believed about disasters that befell an opera company or the habits of hermit crabs, while Larry in a corner taught naughty limericks to a batch of Greeks and Margaret, having fallen for one of Gerry's countless tutors, half died of a broken heart in an attic where she pined and ate huge meals.

Meanwhile the geckos whizzed across lamplit ceilings to enfold a daddy-long-legs in their jaws, dogs like Widdle and Puke infuriated the family at every turn, Quasimodo the musical pigeon waltzed on his perch to Strauss or, to a record by Sousa, marched painstakingly along the tiles. Here too were the mischievous Magenpies, slick and chuckling, resolved to outwit as well as cling to their human environment. Down at the seaside, the clams snapped their mouths shut in a whirl of sand, the crabs adopted a camouflage of anemones, while above the intensely clear sea, in just as clear air, the mantises on glinting foliage were making love, female slowly consuming the male's head as he fertilised her with the remains of his body.

Once Gerry captured a female scorpion, with its carapace infested with young. He placed it in a matchbox, subsequently opened by Larry to light a cigarette, and the resultant chaos at a civilised table sent the entire family into panic retreat. Indeed it was this last incident that convinced Mother that the boy needed the discipline of another tutor. But it was almost too late for the academic to take hold. Durrell had tasted freedom, the joy of wandering alone, making his own finds, a pleasure once tasted never lost. '*Chairete,*' went the Greek greeting, when he met a peasant in the olive groves or a fisherman putting out to sea for the night. Be happy. 'Be happy?' wrote Durrell twenty years later. 'How could one be anything else in such a season?'

The party was about to start. Down the valley I heard a car bumping along the rutted lane. Someone parking shouted briefly. Wasps were buzzing round the necks of wine bottles. Durrell swore under his breath and looked shy and unnerved, a moment of stage fright I had observed before. In rapid succession ten or twelve people appeared in couples, treading the edge of middle age and not specially liking it, cheering up as Gerry opened his arms to administer a number of bear hugs with a broad grin all round. Offering drinks, pouring them, he then settled down into his usual jaunty, bantering relations with close friends, letting his spirits flow freely and the last traces of shyness vanish. There was talk of the quantities of wine we would all soon be putting away, as if no social occasion worked without drink. Opinions were shared about the weather, just to savour aloud the wonder of the day's light and clarity. These exchanges, as always when conducted by Gerry, held out the promise of immeasurable pleasure, a sense of unique occasion.

Someone was asking him about a new book, *The Stationary Ark*, which he was supposed – though it was 'sticking in my craw like a fish-hook' – to be writing here in France. As always when his work was mentioned, a shadow of defensive boredom passed over Durrell's face and he groaned like a fifth-former being dunned for an overdue essay. 'My plan is always to leap out of bed at four and make all my little black marks on white paper between five and nine,' he said, 'thus leaving the rest of the day clear to give parties in. But, ridiculous as time is, before you can say "masterpiece" it's midnight and you're talking balls and the next day's ruined. Oh, I wish I were more disciplined, but on the other hand, if you make it all too rigid, you lose the pleasure.

I suppose, God help me, that this is what my education, if such it can be called, gave me in Corfu: something vaguely based on discipline, but with a wild sense of liberation, rather like riding a horse at the gallop or handling a sailboat, which I also did in Corfu in rather a desultory fashion. Now I wish I'd learned properly, but I was too busy looking at fish and raping peasant girls . . .'

'Actually,' Durrell went on, 'I do wish I'd had a more stringent training in biology, though half the people I know with degrees are so limited – a dogfish only turns into a real dogfish when it's in pieces on the dissecting table. When I was with a dead creature I knew what it was like alive. Although the classroom basics of biology were a closed book to me, walks with Theodore contained discussions of everything from life on Mars to the humblest beetle, and I knew they were all part and parcel, all interlocked. But the young nowadays belong to a standard-work culture, they don't read all the books that are parallel arteries to their own narrow subject, they don't say with a sense of wonder and sheer inquisitiveness, "I wonder what old Darwin thinks about this", and then rush to look this up and fondle his pages. For me in Corfu additional discipline would have made the education perfect. You can't impose your own at that age – the longing to discover, the search for everything under the sun, races out of control like a burning haystack. Live with living things, I say, don't just peer at them in a pool of alcohol.'

Durrell raised his glass to me and half winked. In the background Larry had launched his squat body into a dance, fingers snapping lazily in the air. Greek music pounded out of the tape recorder at full volume. Drawn by the rhythms, Jacquie sketched a few formal steps on the flagstones. 'Greece,' Durrell said, 'has the best popular music in Europe, but it hardly ever crosses the border, probably because they never enter the song contests. I like twingy-twangy music like this, it goes with Vivaldi and the harpsichord. When I was a baby in India the ayah used to start up the gramophone in the morning, no doubt some version of "Come into the garden, Maud", to prevent me from waking up bad-tempered. It might help now. And of course there were those brass bands in Corfu, vying with one another to see who could play the worst.'

Durrell got to his feet and stumbled into position on the terrace and began with unlikely delicacy to toe out the gentle movement of the dance, with upraised arms circling his brother. Mischievously Larry tried to coordinate the ritual twists and turns with kicking his partner's bottom, while he uttered cries, doubtless insulting, in Greek. Once again the moment plunged into a caricature of the old days, a tipsy indulgence in a nostalgia too poignant to admit when sober. Prospero's isle, becalmed in the wine-dark seas of the unthreatened Thirties, went on exerting its spell on the two men, though nowadays they felt very differently about it. Never go back to where you were happy, Larry believed, while Gerry still felt an urge to track down on the island the kind of villa that no longer existed and settle there to recreate times that had gone for ever.

Larry had overcome Mother's initial resistance to the notion of settling for good in Corfu by producing a tutor for Gerry. From nine to lunchtime, George adapted his teaching methods to the pupil's temperament. Arithmetic was arithmetic and largely ignored. Geography meant making multicoloured maps of exotic parts of the world, so that even nowadays Durrell had only to glance at a map to know instantly where he was, what animals to expect, what grew where. Biology meant encouraging the boy to keep a journal, with strict attention to detail, of what he saw and did. English meant reading mainly discoveries of his own – Rabelais (still a top favourite) and the unexpurgated *Lady Chatterley* no less than Fabre, Humboldt, Darwin and 'any such title as *Ninety Years in Baluchistan with Sidelights on Native Characteristics*. I was always puzzled,' Durrell added, 'that he had spent all that time in Baluchistan without allowing anything funny to happen to him. So such humour as I inject into my writing has its roots in those books devoured in the shade of olive trees on hot days – apart from that family characteristic of exaggerating ordinary events into tales of unspeakable hilarity. But I did read everything from *The Soul of the White Ant* to *Three Men in a Boat* via Petronius. I was helped with what I didn't understand or looked it up in the dictionary. There was no question of censorship. Larry used to throw books at me, with a crisp word as to why they were interesting – and I either read them, if he seemed right, or didn't, if lazy.'

The mood of these drowsy lessons was caught by a butterfly, a wily swallowtail of unusual size and beauty, which Gerry could never manage to net for his collection. It made a point of settling for minutes at a time on a lemon tree outside the room where he studied with George. It was the embodiment of freedom, a daily taunt. So a plot was laid with Leslie's help. Whenever Gerry glimpsed the butterfly alighting amid the blossom, he asked to be excused and rushed to alert his brother with the code phrase, 'The old man's in the tree, Les.' Leslie would then dash into the garden and fail to catch it. In the end George realised it was futile to incarcerate for a whole morning a little boy quivering with energy, so he instituted outdoor lessons, geographical strolls to the beach, history in the shade, thus establishing, for the benefit of later tutors, the only way of drumming any information into the head of this very stubborn child. These indolent studies had, needless to say, no formal examination in view. They were beyond time, like everything in the island – including George, suspended in a vague world of his own, 'as people always were on coming to Corfu: their brains went to mush'.

Apart from the family, George was the first person Durrell really studied. 'I used to watch him with great care,' he said. 'After all, I was locked, with him the only specimen, in a kind of human laboratory all morning. And it was the first time I realised that adults could be peculiar – just think how, unobserved, he used to fence with olive trees on the way to the villa and perform cracked little dances to himself. And sometimes he would plunge into such a deep trance at the window that I could stroke a pigeon or stand on my head or even ask questions in a piping voice – and he'd notice nothing. He taught me – what a marvellous lesson! – that people could be eccentric but harmless, odd without being mad. That was a better part of my education than any amount of bookwork craftily aimed at letters after my name. And George wasn't the only one. There was another tutor, who had to go because he fell ponderously in love with Margaret, then there was the brief reign of the Belgian consul who tried to teach me French for no good reason, and finally came dear old Kralefsky, dinning into my head in the old-style Victorian manner all the county towns of England, including lists as long as rows of sub-urban houses of what these towns made or mined or marketed, but such a kindly man saved in the nick of time by his interest in birds from boring me into an early grave. Like everything else in Corfu it was singularly lucky, this string of outlandish professors who taught me nothing that would be remotely useful in making me conform and succeed and flourish, but who gave me the right kind of wealth, who showed me life.'

The guests had eaten and drunk and danced and late in the afternoon were now on their way. Gerry poured more white wine for one or two stragglers who settled here and there in the shade for a siesta long enough to enable them to drive home. Jacquie disappeared indoors and shortly afterwards glasses clinked, plates clattered and a washing-up machine began its low-pitched grumble. If nobody else was paying much attention to practical arrangements, Jacquie knew that they had to leave that night on a three-hour drive north to the Dordogne, where Gerry was due to inspect a privately owned collection of animals which had been offered him for nothing as an adjunct to his own zoo. From the kitchen emerged an occasional sharp reminder of these plans, which Durrell acknowledged with a nod as he winced at the racket from the cutlery. He also ignored the muffled sound of suitcases being packed in another part of the house. The afternoon waned. The guests who had not departed were fast asleep. With a glass of wine before him, Durrell sat idly like a rajah swishing flies off the table with a swatter.

'We used to have a dog called Keeper,' he said reflectively. 'Whenever Jacquie so much as opened a suitcase to start packing, he went into a decline and followed her dismally about the flat, beseeching her not to go away and leave him. Marvellous – if I had a permanent home, I'd fill it with a collection of dogs: a St Bernard, an English sheepdog, a bulldog and a Dalmatian. But that's not to say you should have your life ruled by animals.' He stared at the ground for a moment. Some sizeable ants, abdomens in the air, were making heavy weather of one or two grains of well-spiced rice dropped from someone's fork. 'There were always four dogs in Corfu, for example, not to mention the invasions of peasant hounds in search of a square meal. Like children – like me at the time – any animal needs discipline and should never become a bore. Like children, they're happier with it. When a dog steals a sausage or barks his head off for no reason, it's always the parents' fault, as it were, never the child's, but it's usually the neck of the child you want to wring, which is very unfair. All children are animals at that awful age surrounding puberty, and I was no exception. You could be an animal in Corfu and get away with it.'

He paused. 'Certainly in Corfu I was spoilt,' he said. 'There was a proper degree of licence, but I also came in for a lot of criticism and abuse – in other words, being brought up as an adult and so never really behaving as a child. At least only a fifth of the time, when playing games with the peasant kids – though even games usually led to getting my fingers experimentally on their knicker elastic, which was a fair attempt at an adult joy. Otherwise I loved dinner parties – yes, no less than nowadays. Moths in the lamps, two in the morning, the food gobbled up but the night only just starting, talk of matters I had no means of understanding but found utterly riveting – all very unlike the average child's bored reaction to his elders, if not betters. That's why, when we returned to England, children of my age seemed to me to be confined in such circumscribed lives, existing in cocoons, breathing only when they could snigger over a dog-eared *Health and Efficiency* magazine – which seemed odd after the sheer sexiness of such colour, vibrancy and talk . . .'

Flat on her back near the roses, a woman guest put a hand tenderly to her head. Under an almond tree someone stretched, yawned and staggered to his feet, groaning faintly. The afternoon was waking up. A winged beetle buzzed past, cicadas cranked their endless racket up a notch or two. Indoors a suitcase was slammed shut. 'What we all need is a drink,' Durrell said loudly.

'Oh, do shut up, Gerry!' Jacquie appeared, bowed under the weight of two giant pieces of luggage. 'Everyone's had quite enough, including you.'

'Why didn't you tell me you were doing that?' cried Durrell, at once solicitous, leaping up. Jacquie shrugged and, dumping the suitcases, sloped back into the house. 'Nobody ever tells me anything,' Durrell said, spreading his hands in mock despair. 'No, but what I wanted to say about Corfu is that what I really learned was pleasure.' He lingered on the word. 'Sun and sea. Music. Colours – there's such delight in just looking around a room or a countryside at the colours in it. And textures: rocks, tree barks, the feel of things, almost as though they were edible. Like an oyster being the nearest approach to eating a wave. And I felt this morning that the breakfast honey was the closest you could come to savouring a wet beach – it's how it *should* taste and look. Then books, pictures, any sexual experience. I wasn't brought up a prude, so I see no trace of harm in any activity, provided it gives pleasure to both people and has no baleful influence. Then baths, swimming, water on the body . . . Oh, and colours too are for me linked with the idea of edibility, I could work out colour charts of taste – oysters, for instance, taste green. It all comes, you see, of being a woolly dyed-in-the-wool romantic of the first order and being educated by Corfu into the pleasure of being like that. The island was perfect for it, like a film set, the cypresses stagily against the sky, the olive groves painted on the starry backcloth of the night, a big moon hanging over the water – a Hollywood of the senses arousing in me a shameless and eternal liking for schmaltz. A roaring log fire, three rousing cheers, and bring out your Christmas pudding – I do physically yearn for the cosy, childlike sorts of delights that people like Dickens describe, which is total sentimentality, I suppose. But we had it all in Corfu. And that was my school.'

The last guests were now departing and Durrell saw them off at the gate while Jacquie hurried down to the Mercedes with the suitcases. She checked water, oil, battery, then ran the engine for a few moments, ear cocked for a false note. In her warm, chatty tone she said, 'Of course, Corfu is not good for Gerry. It's almost as if he's fighting against the place that produced such beautiful memories and can't any longer sustain them. He becomes quite intolerable from the moment he sets foot on the quay and realises it will never be what it was, he's emotionally involved in something that can't and won't come up to his standards, and at such times he's just not very nice to know. He hotly denies it, of course. Have you ever known a drunk admit that he's drinking too much? That's why I loathe Corfu – for what it does to him now. Though I do think that he's at last growing out of it.'

Durrell appeared, wearing slacks and a richly patterned Camargue shirt in purple, puffing slightly under the last of the baggage. With evident reluctance he loaded it into the boot. 'Isn't that so, Gerry?' Jacquie said, as if he had overheard.

'What?' he said grumpily.

She explained that our talk had been of Corfu. We climbed into the car. Seat belts clicked. Jacquie backed cautiously into the lane. Durrell wore the closed-in, slightly waxen expression he often assumed when a passenger in his own car. We drove in a state of tension towards the main road norch. We would pick up the motorway outside Orange for the three-hour drive and there was little likelihood of the silence being broken except by snores. The sun was sinking, softening the scrubby hillsides where the gorse grew bottle green. Then to my surprise Durrell said, with a sober formality, as if beginning a lecture, 'Corfu is its coastline, that's all. I am still drawn back to that coastline, but to a lesser degree, because the Greeks have ruined it. When a Greek peasant designs a hotel, he makes the bad taste of a provincial Frenchman look like a stroke of genius. So the coastline is now cut off from the interior of the island, which is virtually untouched, by a welter of chromium plate and glass, by enormous hotels like fifteen cement aircraft hangars joined together – the whole place so delicately painted by Edward Lear now looks clumsier than Clacton.

'Total lack of control, total rapacity, total insensitivity.' His tone relaxed. 'In the old days we used to have – think! – as many as fifty tourists coming ashore every two weeks, producing indescribable chaos and panic on the island. We used to follow their progress by the columns of dust in the distance and rush the other way to avoid them. Nowadays a deserted beach, to the Greek mind, is merely a beach where there are no tourists – but since they're a communal people, it's stuffed with Greeks, kids, screams under the olives and broken bottles. I thought we were safe a few years ago when we found the only remote villa left, with scarcely any approach except from the sea – and from the sea, in their boats, came hordes of huge, sweaty, bristle-headed Huns and awful, languid, vapid Swedes.' Durrell paused. 'And all to what purpose? The visitors don't spend tuppence. They eat in the hotels, they flake and redden on the beach all day, they patronise the town once by buying a couple of postcards. That's their contribution to the economics of the island, otherwise it all goes to tour operators and hotel managers. So Corfu is still poor. And it's ruined. Do you wonder that I'm a conservationist? And to think that up to a few years ago the only place which I would consider home was Corfu.'

'Exactly,' Jacquie said. There was a brief discussion on how much the ruination of Corfu was Durrell's own fault for popularising the paradise by writing best-selling books about it. Then, within moments, his head lolled back into a long-delayed siesta and the virtually silent car murmured up the Rhône valley on the autoroute. The peaches had been harvested from the tidy orchards, grapes were ripening on the battalions of vines that marched upwards into the foothills, the river flowed strongly a few feet from the road, swirling in dangerous currents in the opposite direction to our journey. There was the customary sadness in leaving the south as the sun declined. Dozy myself, sitting in the back, I thought of an earlier retreat. In the spring of 1939, making their way northwards by train from Brindisi, the family had left Corfu: 'heartbroken,' Durrell had said. But even that magic seclusion had been insufficient to dispel rumours of war, at least in the ears of Mother, who was subjected to so much conflicting advice that she ended in a state of panic. Gerry had no idea what war was. 'My entire world was animals. I never looked beyond the boundaries of the island, I lived in a completely selfish cocoon.'

The great blue signs to places I liked or wanted to explore or had stayed in with someone swished by. We were making good time. Durrell recalled that at Lian station the party – four humans, six canaries, four goldfinches, two greenfinches, a linnet, two magpies, a seagull, a pigeon, an owl ('a decent cross-section of birds'), two toads, two tortoises and three dogs – had changed trains. At the Swiss border a bewildered official took one look round the compartment unsteadily piled with vibrating cages, muttered 'one travelling circus with staff', clicked his heels and vanished. 'Some people are peculiar,' said Mother, settling into her corner, little imagining how generously she would have to bribe the staff of their hotel when they at last reached London. On arrival the magpies alone, pecking invisibly at the sacking over their cages, shouting abuse across the foyer in a mixture of Greek and English, convinced the manager that 'we were importing a hitherto unknown species of man-eating vulture'.

We arrived lateish at the well-starred gastronomic hotel which Jacquie had booked by telephone, after an argument sparked by Durrell's fear of being put to bed on sandwiches and tepid lager. When the proprietor, who was also the chef, emerged from the kitchens attired in spotless white, he placed a patronising arm round Durrell's shoulder, as if recognising a trencherman at a glance. Or had he mistaken him, as had sometimes happened, for the long-dead Hemingway? In either case we all washed our hands in a state of scarcely subdued excitement, to be faced at table with a menu of such alarming richness that the very wording seemed to stab at the liver. The hint of a tear appeared in Durrell's eye as he contemplated the menu, then gazed round at the little dining room with carved sideboards, bright linen and a serious array of cutlery. 'Ah,' he said.

Durrell waited in silence until everyone had chosen more or less modestly. He fondled his beard. 'Well,' he said gravely, 'I've no choice but to start with the . . .' Of course he intended to sample every dish the chef deemed worthy of special mention. The *pâté de foie gras* was an obvious choice for this area. You could hardly turn aside an omelette *aux truffes* in a district noted for the finest truffles in France. A few *écrevisses* from the local streams must not be neglected. And since the fellow was patently famed for the way he cooked duck, a duck there must be.

'Oh, Gerry, you can't,' Jacquie said. 'You know what will happen.'

Durrell grinned airily. 'One point I haven't divulged about Corfu,' he said, watching the waiter uncork a bottle of wine. 'I got a fairly bad boil on my side that caused me agony for twenty-four hours, but that was the only unpleasant occurrence during the entire period. Having had chronic catarrh since I was two, propped up with pillows every night to get a wink of sleep, it was wonderful to step into the sun and be instantly cured . . . Of course I was sad when a pet died or disappointed over a picnic not happening. But the only disturbing event of any kind was leaving the island. I remember a writer who was attempting to make a film script of *My Family* saying, "How can you write a film out of this stuff? It's as though it was Christmas every day." Well, in fact, it *was* Christmas every day.'

A substantial wedge of pâté was placed in front of him. 'I say,' Durrell exclaimed, 'that's rather a handsome portion, isn't it? Well – must force it down, I suppose.' He took a big bite. 'Ah,' he muttered, mouth full. 'Pleasure!'

Two hours later I staggered upstairs to my room to digest not only the unpredicted feast but Gerry's story over dinner – despite the ring of truth, was it made up on the spur of the moment? – of how *My Family and Other Animals* came to be written. I pulled the Corfu books out of my suitcase. *My Family* in 1956 had been followed thirteen years later by *Birds, Beasts and Relatives*. The latter was said to contain all the leftover anecdotes so improbable that even Durrell thought no one would believe them. There had been rumours recently of another volume, already written but concealed in a safe, which was to complete the trilogy – to be despatched to the publishers only when Durrell could think of no other book to write but needed the money. This living legend of a possibly nonexistent book was an insurance policy, a best seller under the mattress, no doubt packed with stories that in retrospect Durrell himself, even if he knew they were true, did not quite credit.

Written when Durrell was thirty, *My Family* was a studiously planned success. So at least he claimed. Gerry sometimes pretended to be cunning or devious, whereas more often than not his very innocence of guile in practical matters was a source of astonishment to others. It seemed unlikely, despite his insistence, that he had sat down consciously to manufacture a best seller. So far he had published three animal-collecting books which had plenty of fans. Here surely was the foundation for an audience of millions to rally to the cause. So at least he hoped.

To mix the magic of *My Family*, Durrell maintained, he had started off like a good cook with three ingredients which, delicious alone, were even better in combination: namely, the spellbinding landscape of a Greek island before tourism succeeded in spoiling it for tourists; his discovery of and friendship with the wild denizens, both animal and Greek, of that island; and the eccentric conduct of all members of his family. The first offered escape, sunshine, paradise, peace, which he supposed people wanted; the second, adventure in the natural world, which he knew from experience that people liked; the third, fun, light relief, situation comedy, which he suspected most people found hard to resist.

Durrell thus sat down to juggle, so he said, timing his triple act so that the reader never had a chance to grow tired of one element before the author switched to another. After a riotous session with the family (Larry, for example, finding a beetle in his soup), Durrell would pause and decide whether to write about the breeding habits of the sea horse or describe a sunset turning the sea to burnished gold. The resultant blend had been a best seller at once, scarcely diminishing in sales over twenty years.

I had only to open *My Family* and its sequel at any page to realise – drawn in by the sun's heat, the insect murmur, the unpolluted sea lapping at the deserted beaches – that indeed they offered escape. In first homing in on Corfu the Durrells had been escaping too, from overcast summers and England's all too obvious defects. But the escape had turned into a journey of discovery.

Every member of the family had found his or her version of fulfilment on the island.

Leslie, preoccupied with guns, boats and the chase, fell upon a paradise no less for the sailor than the hunter. Margaret was a physical being who made the odd sortie into arts and crafts, ached for sunshine and sex and a sense of her own beauty; she found all three. For Larry the cleansing light of Greece illuminated the picture of a drab England he produced in *The Black Book*, a young man's masterpiece, but also committed his mind for ever to the Mediterranean. In Corfu he recognised his own voice. And Gerry claimed that for five years of his young life he could recollect the events of any single day. This clarity of recall, sharpened by the elation of discovering nature's infinitude accounted for the almost photographic verve of the wildlife passages in his books. Not only were family dialogues printed on his memory as accurately as on a page, but half-hours with insects popped up in his mind's eye as if already cast in words. It could hardly be otherwise, as you realise when you read. Whether speaking of the oil-beetle larva (hiding in a bloom, hitch-hiking on a bee's back to the cell where the bee laid her egg, which the larva then ate) or young swallows (standing on their heads to deposit pellets of excreta on the nest rim, which the parent bird then dropped elsewhere), the detail is exact, the excitement immense.

When the time came to leave Corfu, Mother gave as her excuse the difficulty of finding another tutor. Gerry was scandalised. He pointed out that, when ignorant, one was much more surprised at everything. The constant simmering of enquiry was at the heart of that extremely funny pair of books about the island. They were themselves an education: which helped to account for the selection of *My Family* as a set book for O-level candidates aged about fifteen. That he should be given at a remove such academic status made the family, including Gerry, laugh with a mixture of scorn and pride. All of which made sense: a book for the child in us, about childhood, by a man who had so little ceased to be a child himself.

Tuesday

EARLY NEXT morning the Durrell bedroom had the busy silence of an intensive-care unit. Hushed conferences on the stairway. Bottles of ice-cold Perrier hurried on trays down corridors. An anxious wait for the local chemist to open.

'I told him so,' said Jacquie. The patient was propped against the pillows, looking stickily pale, quaffing aerated water every few seconds with a dull smack of the lips, and belching. 'This biography,' he said glumly, 'is going to be posthumous.'

There was still time, however, to pursue the day as planned. The idea was to spend an hour inspecting the miniature zoo on offer to Durrell, then begin the long haul north to the coast by autoroute, arriving at Le Havre in time for the night ferry to Southampton. Jacquie never baulked at a tough drive with several hundred miles to cover, and at least Durrell, manoeuvred into the car, could recover gently from last night's excess. In a while he moved slowly downstairs and stood at the bar drinking yet more Perrier from a shaking hand, while waiting for the car to be loaded. 'From India,' Larry had said, 'I recall only a slightly sick creature. We were all brutes, like bears, and we had rough physical shapes that enabled us to be good philistines, but he was always a little bit ailing, fragile and wandlike.'

Although still suffering, for an hour Durrell concentrated his sharp professional attention on the little zoo, an amateur collection given over entirely to monkeys, mostly red howlers from South America. The house was a sombre Renaissance property built round a cobbled courtyard. Uneven floors of dark, polished wood were an estate agent's feature of the deep, well-proportioned rooms. The place felt secret and oppressively hot in its encirclement of tall trees. The environs were honeycombed with prehistoric caves, far enough south for the mutter of cicadas, north enough to feel like an extension of the Home Counties. Thus poised between the moods of two islands, Jersey where the Zoo was and Corfu where the past lived, the place might well have suited the Durrells, who had never owned a home of their own or settled anywhere for long.

Each quietly rejected it on the other's behalf. Her lips pursed, staring hard, Jacquie seemed to be saying that the property lacked the views that Gerry wanted. The message in Durrell's glance was that Jacquie would find it impractical. They had been engaged in this conflict for years. Both were convinced that they were thinking only of the other's best interest. They had bought nothing. They were still of no fixed abode. All they owned was an array of frying pans, books, records, souvenirs and bedclothes. Perhaps their wellbeing depended on a sense of being constantly on the move – though Durrell had recently whispered, as if in confession, that Jacquie really deserved a home of her own by now and he was determined to provide it.

The more the Durrells wandered about this tantalising estate, watching howlers screech and crash in the free-range branches of the beeches, the more crucial seemed this question of a place of their own. Surely there must be a hint of self-destruction in pouring money into cars, aircraft, liners, rents, when what they needed, or liked to long for, or were pretending about, was a spot to live in that was more than a headquarters: their flat in Jersey being just that, an operations room, if not without its luxury, behind the front line of Durrell's battle on behalf of conservation. Every summer in France they spent days searching for properties within a hundred miles of the *mazet* they borrowed from Larry. They were in total agreement about nothing being good enough. At one moment they were to buy land and build the ideal home, but then there was never enough land. It was simpler to convert a ruin, but this would take years. A particular house was perfect, but the view claustrophobic. Another place, though cheap, was still far too costly. Always a fatal flaw gaped, and they retired, almost with relief, from bargains, paradises and white elephants alike, only to start the procedure again next year.

So the verdict was likely to be negative at this hospitable chateau. Jacquie was already showing signs of anxiety to be on the road, while Durrell examined, with the look of a man struggling against nausea, the way the monkeys were being kept, in long grassy cages built out from old stables, and forcing himself to ask questions about their diet.

In the car no one spoke for some time. The mood of relief persisted. Health improved. Was his sickness that morning in part psychosomatic? However desirable the offered zoo, he had no wish to assume more responsibility. His body knew. Now he was looking forward to a stop for coffee. He was humming a tuneless tune. Our spirits lifted. Durrell began to sing. Albeit croaky, this was his response to the open road, even if later he would gaze ahead without words for miles. All who worked with him noted that for hours at a time, when travelling, he stared out at the scenery, appearing not to register a single nuance, or dozed away beautiful weather. 'I always thought it strange,' one of his colleagues had told me, 'sleeping for four hours in the back of a Land-Rover when the whole world is dawning miraculously as if for the last time. But then when he's tuned up, all the cogs working, you realise how very observant he is: seeing a bird everyone else misses, identifying it, comparing it to something silly but absolutely right – a certain stork, say, like a neurotic bishop balancing a cream puff on his head – and you feel ashamed that you haven't noticed these things yourself.'

For the moment, now and then exhaling a sigh, uttering a snort, Durrell was in a daze as if briefly hibernating. Any exclamation about weather or landscape would be answered by a grunt. He had the power to change your mood. Few could change his. After half an hour on the road, up went the weary cry: 'A tiny pint of coffee and a crunchy croissant from a little apple-cheeked old countrywoman – where is she, where is this paragon?' Being the driver, Jacquie was unwilling to stop before making good progress, and so delayed by insisting on the perfect place, thus honouring his wish, but also contriving to speed past one or two perfect places. The quest ended a while later in an anonymous roadside café where, to offset the modesty of the refreshments, everyone made a point of agreeing how clean the lavatories were. One of the pleasures of travelling with the Durrells was the delight

83

they took in glorifying France, at the expense of what Gerry called the old grey island on the far edge of everything nice.

Over the coffee he became carefully talkative, still feeling his way into the morning, about the resilience of nature. I thought at first he meant his own, but his example was the Laysan teal. This modest species of duck was already on the brink of extinction in 1932 when a hurricane wiped out the entire population, except one female sitting on eggs. Nature allowed these eggs to be punctured by curlews and lost. But when the storm died down the surviving teal still had enough living sperm left inside her to fertilise more eggs, with the result that now, half a century later, the world contained over two thousand Laysan teals. Durrell paused in triumph. The world was subtler and more generous than we humans had any right to expect.

'I think a glass of something red pressed to the left kidney,' Durrell said pensively. A mile or two slipped past in silence. As usual he refrained from urging this matter of refreshment, leaving Jacquie in charge of all travel arrangements. In a desultory way he began discussing the advisability of stopping to buy glasses for the consumption of wine in the car (also handy for picnics and useful to have at home), but there was no response from the driver. 'Do you remember the curious case of the dolphin in Pliny?' Durrell said. 'This dolphin used to swim, in the friendly way they have, into a deserted Italian bay, so tame that the village boys could ride on its back. And so its fame spread, and soon parties of people from miles away, tourists if you like, came to gape at the dolphin and stay in the village and drink its wine, and then the elders were so worried that they held a council – and had the dolphin killed . . . Yes, I've got two unarguable reasons for insisting that I've accomplished very little in my life. First of all, the very size and seriousness of the problem is bound to put me in my place – I've been trying to melt the polar icecap with a box of matches. The other point is that when you compare what I've done up to the age of fifty with the achievements at that age of Darwin or Fabre, it shrinks to nothing. It's like offering a green pea to a gorilla with the suggestion it's a square meal. Where the hell did they find the bloody time? In Jersey I go mad trying to locate half an hour to do anything in.'

His voice rose with a self-disgust that sent a chill down the spine. 'I know how limited and false I am,' he said. 'So I'm bound to think that what is credited to me by others is largely untrue. I really am not doing much. But if that's the case with me (and I'm not given to boasting), what about all the others in the same field? How much or how little are *they* achieving beyond all the vaunting and vanity? All English conservationists seem to do is stand around at meetings consulting one another's ties and wondering what school they attended. Question themselves? They don't even know what a question is. They just have all the answers.

'Yet I do question myself,' he went on crossly. 'I suppose self-deprecation helps to keep me sane, if you can call me sane, which you can't. But, yes, I'm lazy. I'm intolerant. And I am timid. So I have to say: if others are pursuing wrongly, as I think they often are, the same aims as myself, how can I of all people expect to do it right? I have to be constantly on my guard against adulation. For a cause like mine it's quite in order to boast a certain amount, yes, but not to believe what you boast yourself. In the end you're taken in by whatever you're shouting, by which time, like all propaganda, it has become quite untrue. What I'm after is truth. No, that's pretentious. Validity. That's what I'm after.'

Durrell said he had been reading, 'with awe, with tears in my eyes', Cecil Woodham-Smith's account of Florence Nightingale. He closely identified with the story because hospitals at that time were exactly as primitive as zoos were nowadays. A gorgeous girl, he said of Florence Nightingale, an appalling woman. 'I couldn't have stood her,' he added, 'for thirty seconds flat.' Then he described the magnitude of mind – 'people haven't got it these days' – which had caused her to insist on stables being fitted with windows because horses too liked seeing out. He plunged into a bitter silence that lasted several miles. Some orchards passed, a few vineyards, a village or two. He bestirred himself. 'Can we stop somewhere and press to our lips, like a nosegay, a pungent little pastis?' Durrell said.

Lunchtime was approaching. Motorway hypnosis had

Jacquie in its thrall, but she also ('though totally different in temperament and outlook, we both like food very much') had an appetite. The unfair view being that French motorway food was a combination of chemicals and vomit, the Mercedes turned off in search of a suitable Routier which would have lashings of vegetable soup, pretty waitresses and 'those cosmetic pears plunged in drink that taste like a Woolworth's shopgirl'.

I consulted the guide. We took a wrong turning. I was properly jeered at for my poor grasp of French topography. We stepped out amid barricades of lorries to find a tiny shed packed with smoke, drink, voices, laughter. Durrell morosely consumed lots of bread to mop up the last traces of his inner disorder. He looked as huge as a lorry driver brooding over the table. He allowed himself a few smiles of satisfaction as a quick procession of soup, pâté, veal, cheese, fruit, reached the table so fast that civilised conversation was impossible. We were all lapped in the noises of quick service and mounting bonhomie.

We tottered out into the diesel fumes of the open air. Within minutes of leaving, Durrell had settled into an upright siesta in the car that lasted for fifty miles. Nobody spoke except in low voices to debate the exact distance between this or that junction and Le Havre. Clouds were now drifting across the sky, at first fairly high and gilded, then low and releasing drizzle. Durrell roused himself. 'Only a few hours from the clutter of retired colonels, the chips as soggy as the landscape,' he said in mock despair, sinking back towards another bout of slumber. 'One day we'll all be intelligent enough to learn that the old country doesn't really exist.'

I sat reflecting on his affectionate contempt for his non-native country, his gloomy view that things never seemed to get better, let alone change. The family's return to the England of 1939 had been no less glum. The dogs were thrust at once into quarantine. The rest of the menagerie was stacked on the top-floor landing at some lodging house in London rented by Mother as a temporary measure. Her heart, as usual, was set on Bournemouth. On returning to this detested soil, Gerry's first act was to buy from a pet shop a marmoset which at once escaped, lolloped downstairs, bit a paying guest and caused pandemonium among the men with newspapers dozing in the lounge. 'The landlady,' Durrell said, 'was torn between a natural British desire to love animals and a natural landlady's desire to prevent anyone doing anything remotely nice.'

The family later moved into a flat in a terraced house off Kensington High Street, not far from the domed Coronet cinema in Notting Hill Gate, which became the educational focus of Gerry's afternoons. On and off Mother drifted down to Bournemouth in search of property while Larry haunted the off-Bloomsbury pubs. Missing the brilliant light of Corfu, Gerry went round the corner alone to catch romance in the dark, Karloff, Laughton, Garbo, launching a lifetime fascination with the cinema.

Bournemouth loomed, as did war. Larry took the risk of returning to Athens, Margaret to Corfu. In short order, as Poland fell and the Germans moved into the Low Countries and France, Margaret managed to find her way home, only to be despatched to South Africa, Larry staying abroad to 'fight the good fight with beer bottles in Cairo', while Leslie, suffering from a perforated eardrum, went to work in munitions. It was decided that Gerry, though 'totally illiterate' in terms of the English school system, must be subjected to a formal education. He and Mother visited a minor public school outside Bournemouth.

This visit was not a success. The headmaster's idea of a suitable test was to invite the candidate, as if imposing a punishment, to write out the Lord's Prayer. Gerry managed the first six words, which had somehow drifted into his pagan memory, and with divine indifference invented the rest. While the headmaster uneasily discussed his academic potential with Mother, he was shown round the labs by the biology master, who at once gained Gerry's confidence by admitting that he had spent a holiday in Greece. 'I didn't quite slap him on the back,' Durrell said, 'but I treated him as I did Theodore and couldn't understand why he was amused.' The teacher reported that young Durrell was both backward and bright (a judgement with which Gerry was never to quarrel). On the way home Mother asked him casually whether, if offered a place, he would like to attend the school. 'Not really,' Gerry said. So with her unerring nose for the eccentric, Mother found another private tutor, a quiet gentleman smelling of eau de Cologne, whose face had been pocked by shrapnel in the First World War.

This clean, neat man had recourse to eau de Cologne as to a drug, popping into the lavatory every so often for a fix. His one academic fault, as Durrell saw it, was teaching him Spanish for a week. To this language he at once responded ('I learned more of it in seven days than of any other subject in fifteen years'), only to be told that Latin would be of more use, a tongue which, apart from making official sense of flora and fauna, dulled him. Otherwise this tutor's influence was good. First he opened Gerry up to libraries. You could stroll down the road and take new worlds home – amazing. You could also demand any book from anywhere in the country – incredible. In Greece, apart from a few heterodox mentors like his brother Larry and Theodore, Gerry had thought he was alone in his researches. Now he belonged to a culture that really knew its stuff and let him share its benefits.

The second gift this tutor brought was words. Larry had intermittently suggested that words had more to recommend them than met the eye. It was obvious to the tutor that Gerry relished words but had no idea how to exploit them: make them say more than they stated, conceal as well as convey emotion, surprise you with a nuance. He had himself written a work on the English poets. His practice was to teach normally for an hour, then slam the textbooks shut in a cloud of dust, pull a volume of verse at random from the shelf and say, 'Come on, let some wind through the brain, dip into this while I go and find you an apple.' In a drift of scent he vanished into the orchard for half an hour, leaving the boy alone with words, English words, well-chosen words in impeccable order. These were moments of revelation he never noticed because he was so absorbed. Years later, when his first book came out, Durrell got a letter from this long-forgotten tutor. He said how embarrassed he felt to have once taught a boy he judged a duckling who had turned out a swan. 'Because of him,' Gerry said.

But there was nobody in Bournemouth to teach Durrell biology. He chose his books at random. He fumbled through the one subject that mattered to him. In theory this was good training, working it all out for himself; in practice it left huge gaps: a self-made man making not quite enough of himself. He needed others to help him find the shortcuts, but who? Even now Durrell was in several minds about the virtues of an education ('if you can dignify it by such a name') to which Corfu had given such a volatile and expansive start. It annoyed him when people wrote out of the blue, assuming that fame conferred limitless leisure on you, to ask for fifteen pages of free information on a species whose habits were recorded in full at the public library down the street. As a nation, he thought, we had lost the gift of doing anything for ourselves. In retrospect it maddened him too that in the only Bournemouth school he ever attended, aged about seven, all the little boys were being stamped out as competitors, while the most important subject was called a nature lesson, given thirty minutes weekly and conducted by the gym mistress. Needless to say he asked this princess of amateurs to tea. Showing her his finds of shells and stones, he looked forward to that half-hour of revelation 'like a drowning man clutching at a straw that existed only in his own mind'.

But that was in the early Thirties. Now in this first year of war the lack of a mentor in the groundwork of biology was more serious. 'What did I miss? Yes, a degree might have helped – but would it? In the long run it might have killed the other side of me. Because of no job, which was because of no degree, only the need to write for a living compelled me to write at all. Also, the degree idea is waved about like a flag to such an extent that one thinks one needs it – when it's only society needing it, and a society I don't much care for. These absolute dolts in my own field simply have the application to store knowledge like a squirrel and regurgitate it all over ruled paper at the right moment. That shows a sense of inferiority on my part, doesn't it? I must say that subsequently there hasn't been much I've been unable to do through lack of a formal education, but that's because I'm brash.'

He laughed. All this talk, flung over the shoulder towards his biographer in the back, had got us round the *périphérique*, avoiding Paris ('What an unforgivable lapse!') and Jacquie was now striking north towards Rouen. Neither the low clouds nor the gloom in the car had lifted much. Boiled sweets were sucked. Once or twice, at intervals almost as long as the gaps between service stations, Durrell mentioned a matter that would soon be requiring a decision at the Zoo. The solitary otter was being despatched to another home in San Diego, where it would enjoy company and possibly breed, so its housing in Jersey must be adapted to lemurs. But of which sort, mongoose or ring-tailed? And it would cost money. To spread the notion of the Trust's real activities – which had nothing to do with amusing visitors by displaying the antics of animals behind bars, but with breeding species on their own terms – it struck Gerry as important that the animal pharmacy, soon to be built, should be visible to the public. But that would also cost money. Her mind on the road, Jacquie vaguely asked where these funds were coming from, and Durrell exploded. 'Look, if I'm slogging my guts out fund-raising in the States, I'm not coming back to a gang of civil servants in my own zoo refusing to spend the money . . .'

Disgruntled silence fell again for a mile or two. But then arose more cheerful talk about the niceties of difference in the smells of various animals. Skunks were like a peasant bus full of garlic, the odour of marmosets resembled stale urine subtly distilled into a form of palatable liqueur. This reminded Durrell that the new marmoset complex at the Zoo was being adopted by a bank. In other words, a commercial institution would be paying regularly for its upkeep. Money again. Might it not be a good idea for the manager to be invited to join the board of trustees and so help raise funds? This positive suggestion, alas, recalled to Durrell the infuriating case of a bank director who, though keen on animals and prepared to be generous, had said, 'Why should I put money into Gerald Durrell's pocket?' 'It's just a convenient way of not giving,' Gerry said. 'People don't listen, they don't read, they don't think. The only way of correcting the impression that I'm a junketing millionaire living off the fat of the animals is by word of mouth and that takes centuries – the Zoo will be bankrupt before the word gets around.'

Various methods of spreading the word were pondered. This book was one, quickly dismissed. Or why not begin with a letter to the editor of the *Jersey Evening Post*? This paper had befriended Gerry in the poor early days, sponsoring an appeal which had saved the Zoo itself from extinction. The theory behind the letter was Jersey's self-interest. The island needed the Zoo, as a tourist attraction, a source of education, a focus of world prestige, but, with honourable exceptions, lifted not a finger to help it. Part of the reason was a popular conviction that Durrell was rich. Would anyone ever believe the truth?

Durrell had no home. He had been married for twenty-five years without buying a house where he could write and live in peace. He had no property or capital to back him. He occupied his flat at the Zoo by courtesy of a trust which, though founded by him, could in theory throw him out. They would be unlikely to do so, for not only were his services to the Zoo too valuable, but at his own request the Trust took everything he gave them for nothing. No director could be more honorary. Only his electricity was free. The normal perks of such an office – telephone, car allowance, petrol – were not at his disposal. He not only claimed no salary, but no serious expenses.

That was to be the first section of his proposed letter to the local paper. So how did he live? Durrell was dependent for income on the sale of his books. He was forced to write for his own survival and for animals whose welfare he thought more urgent than his own. When he made over his personal collection of animals to the Trust as a gift in 1964, it was valued at £15,000. That was enough to buy a more than decent residence somewhere outside Jersey. When he founded the Zoo in the late Fifties, he borrowed £20,000 from his publisher for the purpose, a sum thereafter repaid out of his own pocket or rather by his books. By some quirk of fiscal law he actually lost a further £15,000 on this interest-free loan which was heavily taxed in the United Kingdom.

But think – this letter to the *Post* now getting into its stride – what else Durrell had given, was still giving, for nothing. He had served the Trust as unpaid Director for twelve years, months of which were spent away from Jersey raising funds for the Zoo. No one else could have done it. In commerce or industry in 1975, £10,000 a year would seem a paltry reward for a brilliant man uniquely involved in this work. But Durrell was not receiving £10,000 a year. He earned – this letter, at the risk of boredom, was to repeat – nothing.

A discreet word would have to be added about his personal habits. Jerseymen were not fools. Indeed they were stubborn and hard to persuade. So they were scarcely blind to Durrell's public activities about the island. These proved he wasn't poor. Too often could he be observed enjoying, indeed sharing with others, dozens of oysters, good grouse, great wine. He liked an expensive car or two. But despite his tastes being otherwise simple, his finances were totally insecure. What would happen if the fickle public took against his books? Or if he suddenly lost heart and couldn't write or wanted not to write for a year or two? Or if his doctor told him to take it easy for a few months or a foreign bureaucrat didn't like his face or a country packed with wildlife dissolved into political chaos and he had to cancel one of the expeditions that provided the material for his books? One look at him and you'd never say he was close to the breadline. Yet he always was. Authors always were. They depended on shifts of fashion. They dug their own graves. Only their past was immortal. And the past had no money.

The Zoo – so the peroration ran in this appeal to the local newspaper – was also at all times close to the breadline. To his last breath, contrary to all advice and against his own interest, Durrell was sure to be pouring money and time into this bloody zoo. He wanted to make sure that life in the future . . . It wasn't just for the animals but . . . There was no shortage of arguments to prove him right. It just seemed a pity that Jersey, the whole of Jersey to the last man, woman and child, could not help him a trifle more, if only for the selfish reason that his work brought people from everywhere to the island, transforming it from a tax-evading backwater into an instant cosmopolis of people who cared deeply about life's prospects on this planet. A penny a day? £3.65 a year? Not much. But enough to make all the difference to an institution that by now really belonged to Jersey. Durrell had put it in trust for the island. It was unique in the world.

Gerry rested his case. 'I'm so convinced by it that I doubt if I shall be able to afford dinner tonight,' he said, as Jacquie edged the car past the convoys of massive lorries on the outskirts of Le Havre. The usual nerves about where to eat started showing.

The restaurant chosen had plenty of stars. It turned out to be rather haughty. The overdressed waiters looked with dis-favour on appetite and served hot food with cold insolence. For a while Durrell put up with this resistance to the idea of giving him pleasure. Out of the kitchens a radio station was pumping a diet of pop music interrupted by news items too hurried to be grasped. A complaint was lodged. The head waiter saw to it that the volume was reduced a notch. The Durrells were then treated with complete froideur, as if they might be unable to pay the bill.

It was a relief to drive on to the ferry, even if this meant leaving France. With a determined courtesy Gerry lumbered through the duty-free queues and soon we were settled in the bar.

The vessel shuddered and moved slowly out into the Channel. We sat chatting. Ironic, we thought, that I was asking for more detail about his return from Corfu as a fourteen-year-old and here we were, old friends, crossing the same stretch of water towards the dark England which he preferred to view as a question rather than an answer. To exchange self-mocking glances with Gerry was always a sort of fight: whose humour would survive longer in the stare-out, who would drop his guard first? We drank beer as the ship started to roll slightly, and then he started talking soberly of just this journey back in 1939.

His return to England was a shock. In Greece he had just about discovered puberty, more or less as a delight. Now in an English context came its pains. From these he thought he had never quite recovered. To this day his moral resentment of England had partly to do with her sexual postures and evasions. What he called his sexual career, launched in Corfu with a remarkable purity and lack of inhibition, first took the form of a fumbling but entirely natural investigation of the bodies of the local peasant girls. Off they would ramble, the two of them, up the slopes between the olives, ostensibly to catch butterflies, only to find they had drifted into the myrtle groves. This classical setting was private enough to conceal from casual observation whatever might or might not happen. A slight push, a bit of a game, a grab, darting away, snatching at an excuse for physical contact, tumbling down the hill in a tangle of limbs, panting, abruptly giggling. 'And before you knew where you were,' Durrell said, 'the knickers were off and you were away.'

In England all was not quite so easy and lyrical, not to mention morally permissible. 'I was sixteen and, after examining so many specimens, had a good working knowledge of the subject. All my tutors had taught me sex. It was also discussed quite freely in the house, and I took it for granted that adult women bathed in the nude. So I couldn't understand why in England boys of my own age found something dirty and furtive about it. And I was soon to realise with girlfriends in Bournemouth that I couldn't treat them with quite the same gay abandon in case they thought me naughty and wicked. In some confusion I was forced to retreat to a chaste and stolen kiss on the brow. It was being suddenly flung out of Rabelais into William Morris, from underclothes to curtain materials in one difficult lesson.'

In England, however, was established a pattern which he still found hard to resist, a preference for 'clumping carthorse blondes' with whom he could enjoy give-and-take with no more commitment on either side than a night at the pictures, while his emotional liking was for large, dark eyes and a tiny, shapely figure. 'I always used to go like a bullet for the blondes, but the ones I gave my heart to were the dark ones,' Durrell said. 'My first amour swam into my life when I started dancing lessons in Bournemouth and there she was, looking extraordinarily like Jacquie, and though I longed to drag her into the undergrowth I feared that her fury and lack of response would destroy any chance of seeing her again, which was an unbearable thought. In years I only kissed her once, in a very restrained fashion – William Morris would have been proud of me, as he stamped out yet another butterfly on the wallpaper. But in any case I wasn't too preoccupied with sex beyond a healthy interest, because I had too many other matters to absorb me. I suppose I worked on the good old monastic principle of telling your beads and beating yourself so hard that you forget to masturbate.'

Durrell now claimed that he always practised honesty in his sexual relations. 'I treated women as human beings, which is absolutely fatal,' he said. Corfu had taught him to think this approach natural. Nowadays he used a conscious charm on the opposite sex if he thought they might be of use to the trust, though he never deployed it to private advantage. Suddenly he looked down at himself with a chuckle. Private advantage? He was no longer a latterday Rupert Brooke with a witty tongue, but a hypnotic eye in the frame of a monster. 'Yes, I can make them wriggle just with my eyes now if I want their help, but in those days I think I was handsome, amusing and true to myself.'

At the outset Durrell always made it clear that (a) he wanted to take the girl out, (b) he preferred to make love to her but, if she resisted, he still wanted to take her out, (c) he would never sulk in a corner if the answer was no, (d) he liked her company but had no intention, if she did consent to come to bed, of putting a ring on her finger. And all the girls said (a) three rousing cheers, what a marvellous way to have a relationship and (b) when will you marry me? Durrell's prenuptial days contained a couple of classic instances of 'the impossibility of making a simple point to a woman and persuading her you mean what you say'. Without his knowledge and consent, he claimed, one girl was lured into love by the delicate skill of his flirting. Though she must have inwardly known that the flattery was skin-deep, she so wanted it to be genuine that she let self-deception take her over. The other case was of a recently married woman with a young child. She happened to be staying in Margaret's house in Bournemouth. It seems that in the ecstasy of the moment Durrell uttered such superficial endearments as 'I can't live without you'. The next morning he was astonished to find Margaret accusing him of 'playing around' simply because the deluded newlywed, whose husband was due home at any minute, was packing her bags and talking of the new life in Devon which she and Gerry would be forging together. 'I've always had a fair amount of attraction for women,' he said, 'but I hope I've never used it — except to seduce them, of course. Their attentions always flatter me too, but I've no need of it, I wouldn't worry if nothing happened. What does worry me is the way you can twist them round any of your fingers, let alone the little one. I could have lived in luxury as a parasite twenty times over just because women force themselves into believing things which inside they know to be totally untrue.'

Durrell stood up from his empty glass and moved in a stately manner towards the decks. The lights of France were receding. England lay unseen ahead. He took several deep breaths at the rail, as if beginning a yoga session, and chatted about the innocence, self-delusion and lack of care for others which inevitably we all brought to our first sexual encounters. Sex was like learning to breathe when you already thought you could. It had about as many ethical implications as an amoeba dividing, except those we imposed because trained by our brand of civilisation to suppose it a moral area. Gerry seemed to have got closer than most to ridding sex of its shibboleths. He had behaved naturally. He had learned from it. He treated women with the care and detachment he would accord a different species, but also with the affection. France vanished into the mists. With a sigh Durrell returned to the comfort of the bar downstairs.

'Then — jumping ahead a bit — I made the fatal mistake of going up to London and meeting my first adult,' Durrell said. 'Larry was ensconced in the Hog in the Pound with a mass of failed poets clustered round him like iron filings and we all went exuberantly off to lunch, including this girl. Girl? No, woman. She was thirtyish, and me ten years younger, lovely eyes but not at all good-looking, just one of those faces that haunt you (so many people have unmemorable faces), and she was pale and looked very sad, and she was out of the Larry circle so I made an effort, and found she painted horses for a living and had abandoned her husband and had two children, and because I had no idea that anyone actually *did* any of these things, I went to look at her pictures and her kids and herself, and I liked it all – she wasn't as good as Stubbs but damned near it – so we went out to dinner and I moved in, virtually that night, give or take a suitcase, and within days installed my trunks of books in her cellar.

'The children took to me in a big way and I whizzed them off to Kew Gardens and the river. I always seem to survive with kids because we're on the same mental level, they always enjoy things I enjoy doing, like climbing trees and falling into duck ponds – at least they love it physically, I relish it more on a spiritual plane. So I thought it would last, which would have proved a disaster, and I would marry her, which she was too intelligent to permit. And here's another little sidelight on the nature of women – we dined one night, I'm not sure we didn't go to the cinema, we went to bed, all as calm and soothing as ever, and the next morning I toddled off to Bournemouth for some reason, and that night I returned to find on the dressing table (where else?) a note in the traditional manner, using all the trite phrases out of the novels we secretly enjoy but pretend are bad, like "we're growing too fond of each other" and this sort of crap, instead of just saying "I'm getting increasingly bored with you, why don't you sod off?", and I went absolutely mad – got as drunk as a newt – the most infuriating point being that she'd taken the electric blanket with her, and I phoned every conceivable place, shooting up her bill to somewhere near the two hundred quid mark, and no luck. But at last a friend of hers came and discovered me in a broken state. So told her. So she phoned and tore me off a strip, which had the desired effect of making me angry – I couldn't wait to get my books out of her cellar. But by no flicker of the eyelid, when we were so happily together, had she suggested any such treachery. Women? I prefer animals.'

It was past midnight and in mid-Channel the bar was closing. By an exercise of charm Durrell managed to secure another beer, which he sat meditatively sipping. 'There has been no agony about it,' Durrell said, almost with regret. 'All I've experienced are the pleasant pains of calf-love, arising out of an inability to get so-and-so to bed, which used to give me that agreeable sensation of wanting to rush out and hang myself in the barn. There's a lot to be said for missing another human being, if you can make yourself aware of the joy of so doing. I would never put off an animal expedition or some unrepeatable event (for example, a show of horse painting) for a girl – she would take second place and like or lump it. But I never mucked them up; in fact, I went to considerable lengths to ensure that it was fun for both of us. There was always respect and generosity. You see, I never wanted to batten on to women.'

He grinned as if to cast doubt on this disingenuous state-ment. 'No, I mean, any money I had was spent on them with great lavishness. Rich or poor, I'm never mean, which tells you only that I have no sense of cash, I don't know what money is unless for some reason I haven't got any. One of the nice things about earning the stuff is that you can contrive for yourself so much pleasure by giving it away.

'But that's beside the point. I'm trying to pin down this really deep-dyed selfishness of mine which, by a happy chance, looks to everyone else like a nice, friendly, generous quality. For example, I'm not interested in sleeping with anyone else. It only even occurs to me to be unfaithful when Jacquie gets on my wick (which of course she frequently does, being Jacquie, being a wife, being in fact a woman), but I don't regard that as a virtue neces-sarily. What's the point of cluttering up your life with somebody else, if that's not the person with whom you want to do practically everything? You see? Purely selfish. I see lots of girls whom I think it would be nice to take to bed, but it's just a harmless fancy, a moment of appreciation, casting an eye on a pretty picture but not wanting to have it on the wall.

'Funnily enough, when it comes to flirting with people and manipulating them, I always enjoy some success with a certain unlikely type of man. The big and beefy sort. Outdoor, cricket-playing, rugger-loving. The heartier he is, the more deeply he falls in love with me – I suppose they're men who would faint or knock you down if you applied a word like "queer", but whose homosexuality is much nearer the surface than they think. I know, thank God, that I've got a strong feminine streak in me, I wouldn't be without it for anything, but I've always been committed to women, not men – not even as friends. There are certain men I can spend a certain time with and not get bored, but women amuse me constantly, it's like dancing through life, but in no derogatory sense, with a flock of butterflies. Their minds are so much quicker than men's. In eight cases out of ten they win by intuition. The man reaches his conclusion by a ponderous ava-lanche of logic while the woman got to the point three hours ago. I'm very slow in some ways, but the feminine side of my mind keeps me four jumps ahead, even when I don't appear to be. In any case there's no choice if you're married to Jacquie. She's so quick she stimulates the mind like an extra supply of adrenalin.'

His last beer finished, Durrell glanced coldly at the bar, now closed, and stood up without a word to make his way to the cabins. The ship was shuddering in the grip of a mid-Channel swell. People shrouded in overcoats dozed uneasily on the bar couches. 'I do like women,' he murmured. 'If they make up their minds to come, for example, on an expedition, they don't moan about the heat and flies to half the extent men do, they don't screech and carry on.' He turned into the narrow corridor leading to his cabin. 'I've always been a woman's libber,' he said, 'but I've no patience with the self-indulgence of the bra-burning stupidity or this awful business of sitting bowler-hatted in front of a mirror wishing you could grow a walrus moustache. But crushed by men they certainly are. They're still groomed for marriage like stud cattle to achieve the extraordinary goal of getting a little band round their finger – it's as appalling as slavery – whereupon the husband gets a licence to beat them to a jelly, screw them stupid and chain them to a stove. In this respect Mother did fight a Victorian rearguard action, but imagine the arrogance of Larry at twenty! She just hadn't a hope. How Margaret remained a virgin so long, I'll never know, and neither does she. I suppose Mother must have been basically one of us or she wouldn't have given in so easily. She worked on the moral principle that, provided your activities did no harm to others, it was probably all right, dear. She would have accepted any act of God, three raging queers for sons and a prostitute daughter, in an all-embracing Christian fashion – so she was really quite happy at the way things turned out. It hasn't been bad, has it? Goodnight.'

Wednesday

THE FERRY docked at Southampton shortly before seven the next morning. A sluggish dawn suffused the New Forest, ponies vanishing in a swirl of mist. On the road to Bournemouth, Durrell stared at the mauve acres of heather leading to fringes of trees at the horizon. He was anxious to point out landmarks: lanes he had cycled down during the war; tracts across which he had ridden on horseback when working for a stables; Christchurch Priory, from the tower of which he had surveyed on high the landscape of his childhood. 'Margaret won't be up, of course,' he warned as the car turned in and out of run-down, respectable streets to the north of the town, stopping at length outside a semidetached that looked like a refugee from the London suburbs taking the sea air. Here Durrell had assembled, before he found and moved to Jersey, a menagerie that kept getting out of hand, to the consternation of neighbours.

His sister Margaret was, of course, not up — it was only eight o'clock — but she appeared for a few confused seconds, bonny, blonde, vague, rather ghostly in a white nightdress, and then, after a swift interchange of arrangements, vanished indoors. 'Typical,' Durrell said with that teasing pride he reserved for members of the family. 'She probably had some scrounging Greek sailor in her bed.' This was a reference to her recent stint as a shopkeeper on Greek cruises. For some time she had been threatening, as if to shame her literary brothers, to write a frank and funny book about the experience.

After driving into the centre of Bournemouth, past clumps of pines in the salubrious parks, the Durrells checked into a hotel with sea views where they had booked a room for the day. Durrell took off his jacket, let out a great sigh, suggested tea and toast, and stepped out on the balcony, where, like a captain on his bridge, he surveyed the spread of the bay. Swanage crouched on the horizon, the swell of the Purbeck Hills mysterious in the background: a view from childhood. He had to admit, though he rarely did, that he felt at home here. Too many memories crowded out his scorn of coastal suburbia, though he snatched any chance of comparing it unfavourably with France, Argentina and West Africa. The hotel cooperated in this mood. No tea or toast arrived. He was saying this was typical of Bournemouth when a foreign waiter appeared with tea and toast and Durrell motioned someone else – he never carried cash – to give the man a tip. He lay on his bed greedily slurping tea.

We were all supposed to meet Margaret in a certain pub at eleven. Durrell thought she had been too sleepy to take in this arrangement. That would be typical too. A message should therefore be left at wherever in town – bars, a reception desk, the pierhead – Margaret might think she had been asked to come. All this Durrell treated partly as a joke, but with tenderness in his tone. Jacquie said that she, whatever anyone else intended to do, was going shopping – no, not for herself (that also was 'typical') but for presents. Meanwhile, she added to Gerry's astonishment, why didn't the men go down to the lounge and have some beer, since this was plainly the way his mind was moving? 'What a very good idea!' said Gerry, leaping off the bed with a groan. 'I've done no yoga this morning and, darling, where are my shoes?' It was not quite ten o'clock.

In the lounge Durrell pressed a bell. At a distant table two businessmen were shuffling papers and conversing in a drab murmur. An elderly lady drifted past, pouting a withered smile at anyone in sight. No waitress appeared. Durrell rang again. At once a foreign girl in black stood staring at him. 'Coffee,' she said thickly. 'Beer,' he replied. Both sounded like native greetings. The girl gulped and withdrew. In ten minutes she returned to tell him that the bar was not open. Patiently Durrell informed her that even in England guests in hotels could by law obtain ale when they chose. Through his beard he offered her an enchanting smile, of the sort known to wow women.

For a space, serious talk atrophied. Durrell was content with muttering about the England on display in this moribund lounge. 'Typical,' he again said more than once. 'Can't you see why I refuse to live here?' With a showy jangling of keys, the bar trellis was slid back just far enough to release a couple of light ales. Incorrectly poured, they were surmounted by lukewarm froth. The girl in black set them on the table with pursed lips. They were expensive. But a battle had been won. 'Bournemouth has at last opened the floodgates,' Durrell said. 'I never thought it would happen.'

He drank more deeply and sighed. It was always hard to imagine anyone enjoying a drink more than Gerry did – tea, tonic water, wine, Scotch or tepid ale. Beyond the picture window a leaden patch of sea was lit briefly by a shaft of sunlight. The curtains framing it hung heavy. 'Are you surprised,' Durrell said, 'that when the Germans by mistake dropped a couple of bombs on this place, all the retired colonels wrote letters of protest to the local paper about it?'

He paused. 'They never knew what war was,' he said, 'but then nor did I, spoilt as usual. We used to see Southampton get a pasting, eagerly enjoying the eastern sky aflame, and there were plenty of jolly dogfights upstairs – but on the whole we had a cushy war. The entire family did. We were pinned to the nine-o'clock news, cheering for a victory, and I followed daily the progress of the battles on whatever front it was . . . but only selfishly. I wanted to get the war over as fast as possible and do something interesting, like return to Greece and see how the Germans had behaved to the swallowtails and trapdoor spiders. Even so, I spent every spare moment out of doors – aged fifteen to twenty – risking death at the hands of the bombers on the way to Coventry or somewhere that really copped it. I helped with the harvest. I went out – not on a donkey now, but a bicycle, not so stimulating, but less recalcitrant – looking for nests and animals, rediscovering the local fauna with more patience and a maturer knowledge, like waiting for the bird to return to her nest to make sure of the species, at any hour of day or night, because I was used in Corfu to regarding the villa merely as a dormitory. The outside was home.'

Brow furrowed, he looked back on that earlier Bourne-mouth. 'Behind the houses opposite,' he continued, 'there was the tail end of a grand estate fallen on bad times and probably sold to speculators, but nobody was building so it remained a dense woodland where in the glades wild strawberries grew and you could still find many species that echoed Corfu, which was similar in its wildlife, though more exotic. Thus two hundred yards from the house I had the woods to keep an eye on, and then at the end of the road golf links, beyond which the country started. The real country. Bournemouth in my time was a country town, not this orgy of mismanaged overdevelopment designed for prison sentences. It was ideal from my point of view – though of course not at all ideal, because it wasn't Corfu.'

In 1942, however, he got his papers to register for military service. A single thought, apart from avoiding war totally if possible, was in Durrell's mind at his medical in Southampton: how to produce a urine sample to order. Having drunk a lot of tea in the train, he urinated so generously into the beaker that it slopped over the doctor's hand. This was his first encounter with the state. In the intelligence test he spent so much time helping an illiterate to understand the instructions that he failed his own paper. In the eyesight examination he interpreted so literally the doctor's order to follow his finger as it moved away from him that he lunged across the room after it. When asked for a preference he put down the Veterinary Corps – 'given a chance of how you wanted to be killed, I thought I might as well die happy' – which the clerk was unable to locate in the handbook. Finally he helped the doctor by diagnosing his own sinus trouble, to be sharply told to mind his own business. Found to be correct, he was sent to a specialist adept in the detection of lead-swinging, a burly man with an eye for cowardice. Durrell's answer to his question about whether he wanted to fight – 'No, of course I don't, sir, but I will if I must' – coincided with this doctor's view that even a serious operation was unlikely to work, and Durrell heard no more from the War Office. 'They had just passed as A1 three men who promptly died of pneumonia,' he explained.

For him, coming to terms with the war also excluded work in a factory or labouring on the land. But just north of Bournemouth he found a riding school which kept a few cows. A bargain was struck. If he mucked out and groomed the twenty-two horses and conducted people on half a dozen rides a day, the owner would tell the authorities he was running a farm. Unpatriotic? And selfish? Durrell had never had reason to suppose that England was his country, even by adoption. In India, then Greece, the freedom of the world had been granted him early. From the start he thought that no country had a natural or god-given right to win or survive against any other. Politics killed people. They were a cruel game. In youth, as if propelled by unconscious foreknowledge, he was preparing for work that had to do with the longer-term wellbeing of the world.

Leaving a further message for Margaret at the reception desk, Durrell strolled past the Pavilion through public gardens which in his younger days contained one of the few remaining colonies of red squirrels in England. Elderly men sat on benches, staring into space, pigeons pecking round their feet. The bandstand was empty. Still brooding on the war, Durrell suddenly said, 'I was at a dance once and heard through the music the rumble of bombs or guns which seemed in the direction of our house, so I thought I'd better give a tinkle to the old mater. No reply – it never occurred to me they'd be in the shelter – but without a word from me the operator said, "Don't worry, sir, it's not the Queen's Park area at all, your family will be all right." A tiny example of the wartime spirit that everyone talks about and of what socialism should ideally be and never is – it's so overburdened with bureaucracy that nobody knows what anybody's doing and so they lose all sense of purpose, even of life. Look at these people here, half-dead memorials to socialism and all of them longing for a war. The besetting sin of our society is that without a war nobody believes there's any danger – and there is, by God there is. For blood, sweat and tears, now read overpopulation, ruin, pollution. For toil read oil. In any case every war is fought on biological grounds, another reason why people should put biology at the top of their reading list. Once it was Lebensraum, now it's oil – since we insanely geared our entire economies to this ridiculous substance.'

One or two passers-by eyed Durrell with caution, perhaps half recognising him. 'It's all to cock,' he snapped. 'In earlier times wars kept a certain balance. In the tribal system two adjacent villages marked out their hunting areas like animals and kept a no-man's-land in between. It's as though life had given us a delicately adjusted clock to tell us the time for ever, and without knowing how the hell it works we at once open up the back and start fiddling around with a blunt screwdriver. Though smaller units are always desirable, they won't work for us any more – the clock's gone haywire. The only way for us to survive in Europe, faced with a banjo-twanging monolith on the other side of the Atlantic and a melancholy mammoth brooding across Asia, is to achieve unity. It's no longer possible to survive at all except by providing just enough to get by on – and even that's dependent on the not far distant day when they bring in compulsory birth control.'

Durrell emerged from the gardens and, dodging traffic, crossed the square at an unorthodox angle. 'The duck-brained liberals and the Catholics – represented by that throaty voice encrusted with the dust of ages that occasionally creaks forth from the depths of the Vatican – say that we must simply redistribute food.' He sidestepped a hooting car. 'But the only effect of that is to save the ones who are now dying, which in turn only results in more population. Someone gets a meal, survives for another twenty-four hours, celebrates you know how and creates one more child. Sensible conferences on the subject invariably descend into senseless politics. Why shouldn't we breed? say the blacks, blinding themselves to the simple equation – copulation plus conception equals starvation. Ah, here we are at last. I can guarantee Margaret won't have turned up.'

It was a saloon bar typical of old Bournemouth, well preserved for its age, a lot of leather upholstery reflected in engraved mirrors. No one had arrived. Durrell sniffed, as if scenting home ground. Stale beer and yesterday's cigars hung on the air, a hint of polish. As ever loyal to the drink of the country, he ordered a pint of bitter. This was the pub he had used towards the end of the war if he could afford it when laying his postwar plans, knowing already that he was to become an animal collector.

He had chosen his profession years earlier in Corfu. It

sprang from a literary experience. Among the packages of papers from England that made up the family's bedside and lavatory reading – *Mirror, New Statesman, Times Lit Supp*, fashion mags, comics – was *Wide World*, to which Leslie subscribed. It serialised an account by Ivan Sanderson of the Percy Sladen expedition to the Cameroons. This dangerous squib fell into Gerry's hands, exploding for ever any notion of easier or better-paid professions. It was the first book of travel and adventure, out of whole dull libraries he had consumed in the shade on brilliant afternoons, that displayed a sense of humour in the writing. He lapped it up. That flicker of fun in and out of the prose was the final push he needed. At once he made a vow to become a collector of live animals, preferring them to dead ones ('not that I subscribe to any shibboleths about killing animals'), and to collect live in the Cameroons all the creatures that Sladen had trapped, shot, stuffed and pickled. 'And by God I did, almost all of them. You see, nothing has ever deflected me. I'm just sheer bloody stubborn.'

In France the Allies advanced as Durrell's interest in the war retreated, victory plainly inevitable. He continued working daily at the stables. By now he had his own horse. On his day off he trotted for hours in the pine forest down the straight sunlit rides. On workdays beautiful girls turned up for instruction. They not only showed promise on horseback, but were gladly swept away across the heather for extramural studies: 'with packed goodies in the saddlebag and time ahead of us, I gave her a few titillating riding lessons, raced her, made her jump and returned happy at the end of the day'. In these escapes or escapades he owed less to amorous technique, he said, than to the magic of the surroundings. The mood of those woods was so otherworldly that his victims always behaved like women on shipboard. In such offshore romances, however many husbands they had, morals did not count.

When Durrell was alone, his rapport with the horse was unthinking. He could sit and meditate on it. He made up poetry in the saddle. He stopped at a certain pub for a quick pint. The horse too formed habits. Durrell later discovered that when the owner rode her, she refused to budge at the pub in question until he had a drink.

Meanwhile he had his animals at home to think of, not to mention his relatives: the Durrell formula went through its difficult adolescence in outer Bournemouth. Geckos in biscuit tins or baby hedgehogs in bed not only strained the family's patience, they were also inconvenient pets, and they could advance him no further on the path everyone else regarded as insane. To collect animals for a living he needed experience, a high class and intensity of experience. This started with such basics as 'how to get a stranglehold on a giraffe or sidestep a charging tiger'. More immediately, but with less drama, it was how to clean out the tiger and what to feed the giraffe. Therefore, aged nineteen, Gerry put his qualifications at the disposal of the London Zoo and sat back to await a telegram of welcome.

He was offered a post as student keeper at Whipsnade. This exotic spot in the Home Counties kept him for eighteen months. Thirty years later he wrote (in *A Bevy of Beasts*) about how much he learned from tending very different animals of more or less unpredictable habits from equally unknown climes. Whipsnade was his university. He looked after bears, lions, gazelles, wolves. From their ways he began to draw conclusions – and, being Durrell, drew them swiftly. His warm feelings about animals hardened into the cold theory of what he must do. The heart of it came to him one day when bottle-feeding a baby Père David deer. This was a creature that reached the verge of extinction in the imperial gardens of China, only to be snatched back in the nick of time by human effort. He realised then what a 'rare' animal was: a species at death's door. As he started compiling his own list of such creatures – rhinos numbered in dozens, Arabian oryx down to thirty or so, the flightless rail down to the size of a small English hamlet or a pub's population at closing time – he felt a mixture of horror, despair, determination and love. So little done, so much to do, as Rhodes put it. Or Chaucer: the lyf so short, the crafte so long to lerne. It was then Gerry reached two conclusions that were to govern his life. First, any zoo he started in the future must 'act as a reservoir and sanctuary for those harried creatures'. Next, any animals close to dying out, faced by human indifference, had as much right to existence as he did himself.

In lesser ways his life as a student keeper was not so fulfilling. Gerry grinned. 'The trouble with Whipsnade,' he said, 'is that it's so far off the beaten track that it was useless pursuing the pretty girls who arrived for the day by the busload unless you could make a really rapid killing – and I was so busy cleaning out cages and learning what to look for in the excreta of monkeys that I hadn't even time to get a hand to their knees, let alone identify the shade of their knickers.' But the Durrell luck, the luck to which he always attributed any kind of success, turned. It so happened that a policeman's daughter, buxom, blonde and willing, was in charge of pets' corner, that harmless enclave of animals designed for the entertainment of children. Her head-quarters was a wooden hut. Its door at lunchtime was often locked from the inside. 'Gone to earth,' said Durrell.

He fetched another couple of pints. 'This is typical of Margaret,' he said. 'We'll be drunk before she appears. I wonder where Jacquie is.' He began to fret slightly. Nobody was behaving according to plan. Perhaps there had been an accident. He sighed. 'I thought I would marry in the end,' he said. 'I had no thoughts against marriage. It's just that so far I didn't want to marry any particular person. When I had an affair with a girl, I could predict with absolute accuracy the other affair that either she or I would shortly be having. Even this girl at Whipsnade, who was super, had some shadowy boyfriend who kept asking her – rather perversely, I thought – if I had been daring enough to actually try kissing her, and she was always giggling about this, even more perverse since we were poking the living daylights out of each other at every opportunity. No, there's only one person I've met whom I could tolerate for any length of time,' he said, 'and that's Jacquie. I wonder where she is? I say, this beer's good, let's have some more. What time is it? I knew Margaret would mess up the whole arrangement. When I first met Jacquie it took me about a month to decide – not altogether surprising. You don't find it easy to reach conclusions when someone fights you off tooth and nail.'

111

He stopped, rubbed his eyes and grinned anxiously at me. It was well past eleven, the hour for meeting. Where were these women? At this moment, loaded with parcels, Jacquie entered the bar. With a brief defensive look at Durrell, calculating his mood, she asked where Margaret was, as if it were his fault she had failed to turn up.

'Where have you been?' said Durrell.

'I told you, I've been shopping,' Jacquie replied, shaping up to a quarrel. 'I suppose you men are pissed out of your minds.'

Margaret promptly appeared, dressed in white as for a wedding, with a wide-eyed look of amazement that the Durrells should actually be in the agreed place. There was a strong family resemblance not only in her features – wide mouth, direct and penetrating eyes of a strong blue – but also in her smiling, slightly metallic, mock-innocent tone of voice. Durrell was courtly with his sister, standing up, waving her to a chair, fussing over drinks – his manners were often a striking source of pleasure to anyone in his company – and the next half-hour passed in discussion of immediate plans. Margaret was flying over to Jersey at the weekend for the Annual General Meeting of the Trust. Jacquie had to accompany the car on the ferry from Weymouth; they had come this roundabout way to Jersey only because the direct service from France was suspended. To avoid delaying Durrell, who needed time to prepare for the festal climax to the Zoo's year, but who detested normal flying, a six-seater aircraft had been hired to take him across the Channel later that afternoon. Durrell was never more spirited than when plotting the future. Here in this pub he had the world at his feet, to be altered at will by his choices. It was rather like the time in Bournemouth when at twenty-one he decided to blow the £3,000 of his inheritance on the trip to the rainforests of the Cameroons that had tempted him seven years earlier in the pages of *Wide World* in Corfu.

This was in 1947. Durrell told the story of that expedition six years later in *The Overloaded Ark*, his first book, written mostly because after two more trips, to Guiana and Bafut, he was flat broke, living back in Bournemouth on ten Woodbines a day and much tea. In 1949 he had met Jacquie in a Manchester boarding house. In 1951 they eloped southwards, married on a few pounds, and took up residence in a tiny, self-contained corner of Margaret's house. He had three successful expeditions under his belt, and an allowance of three pounds a week from his mother. He began writing.

'It all started long before, when Leslie used to come home from Dulwich on holiday and tell me Billy Bunter stories,' Durrell said. 'He used to embellish them with his own bits and pieces, add a dash of his own school adventures, imitate a master or two in a very clever and vivid fashion. He had the same gift as Larry, only untutored, not so well developed, and unconsciously I must have been absorbing the fact that this was the way to tell a story. Because when I started writing *The Overloaded Ark* I found it difficult to convey a character until I discovered how to do it by description and a trick of speech – and most people have tricks of speech. My first impulse was to imitate physically what I was trying to get over to the reader, but you can't imitate on paper, so I had to sit down and learn the knack of translating an imitation into words by means of timing and suitable exaggeration and knowing how, in the film sense, to cut. It's an untidy book, *The Overloaded Ark*, but I was learning to edit events so that they offset one another funnily, highlighting an episode that may have happened at the tail end of an expedition by twisting the whole plot to put it up front.'

Margaret had taken over direction of the house from Mother, who thus became an honoured tenant. And she threw it open to everyone. 'It was normal practice,' Margaret said, 'for people just to turn up with friends and stay for half an hour or a month. Gerry had no roots at that time, so turned to me – any letter with a foreign stamp was an ominous sign of things to come. Suddenly a van would drive up and land a crate of pelicans on the doorstep and you'd be rushing about trying to find fish and you knew that in the horrible not-so-far-away Gerry's return was imminent. Up to the time they went to settle in Jersey, there were always some animals in the house and everyone, family, friends, lodgers, lent a hand in looking after them. I was always penniless, so lived by letting the house, and for the ten years after 1948 there was always room for Gerry and Jacquie.'

In the suburban street, there were inevitable complaints. The snakes, apes and toucans caused protest. Marmosets and spiders were not popular. A neighbour who kept chickens angrily summoned the public-health authorities to inspect the fleas escaping from the menagerie next door. On examination they turned out to be chicken fleas. The local sanitary inspector grew so tired of being called in to condemn the lack of hygiene in the back-yard zoo that soon, eyes glazed, without a glance at the animals, he was giving them a clean bill of health. On one side lived a retired engine driver whose wife aired his overalls on the garden line. On the other, a retired Indian High Court judge and his lady were too refined to complain beyond an occasional whisper about noise. 'We evidently weren't a "common" bunch,' Margaret said, 'so Lady Thing didn't know how to cope with us. When a monkey knocked over a lamp in her drawing room, all she did was write us an apologetic note – though she did have a vapour when a carcase was carried into the house and Gerry set up a chopping block out front by the garage. Not everyone enjoys lifting a cushion and finding a reptile under it, but our lodgers soon got into the spirit of the thing.' She paused, then added, 'And there I was always trying to sort out some respectability for myself.'

Meanwhile the retired railwayman was buying up as investment the pine woods behind their houses. 'Now they're a wealthy family,' Margaret said, 'and I'm still struggling. Mother was always too cautious and I never had money or looked ahead. All we Durrells have this streak of preferring to be busy enjoying things rather than getting down to a spot of hard work. When Gerry and Jacquie were there, it was all tea-drinking and chatter, enlivened by excursions to see other people's zoos and inspect places for his own.

'Yes, I do think Gerry has changed since those days,' she went on. 'He would be the first to admit that he had a much greater sense of humour at that time, almost as if success had embittered him. He was never worried about money, more in touch with ordinary people, more tolerant, and his zest for living was higher – those trips looking for property were a riot of rude songs and Greek choruses. But just at the point when the family became less closely knit, other things began crowding into his life. He wants and expects everything to be idyllic, but that's not possible when big business and bankers are involved – he would be a much happier man with a small zoo that somehow financed itself.'

She thought for a moment. 'The trouble when you're famous,' she said, 'is that you're always performing, never yourself – until in the end you forget how to be yourself. Provided I've got my beauty aids, a toothbrush and a pair of shoes I can walk in, I can go anywhere and do anything. Food and drink aren't important to me, I'm free. But with Gerry, people have come to expect clever remarks and witty anecdotes, and then the tendency is to overuse yourself, which means that the essence of yourself gets lost. It's not the same with Larry. He's a clown anyway. But the whole family possesses a quality, or menace, of unpredictable exuberance that will carry people with you. It comes and goes, like falling in love. One wonders if one is ever going to feel it again, and then there it suddenly is.'

She was speaking directly. She seemed not to care whether her brother overheard these strictures. At one level all Durrells were optimists. Soon, his financial anxieties at an end, no longer exploited by good causes, he would revert to the high old humour of the good old days. Meanwhile, against his nature, he had forced himself into the public eye to save the lives of animals who had all too few protectors. It wasn't surprising he had changed. But there were always moments. Like now. We had left the pub and Gerry was briskly joking with the taxi driver taking us to lunch. To everyone's apparent delight he was reciting doggerel by Cumberland Clark, the Bournemouth poetaster who had punctuated Gerry's twenties with bad rhymes and even worse sentiments. According to Margaret, this was the high-spirited, soft-natured youth of the Bournemouth years taking over, shaking sense into the middle-aged man who had become, in the family view, solemn and grumpy. Jacquie seemed to share this view of him. She thought those penniless days in the one-room flat at the back of town were in many ways the best they had known.

In the redbrick station hotel on a bleak corner, Gerry stood in the lobby, puffing on a large, fat cigar made in Costa Rica. The restaurant had long menus and was busy with businessmen at lunchtime. Durrell strode off to the phone to communicate the agreed arrangements to Jersey, making sure of his reception committee. Meanwhile Jacquie took up the story. 'It was a bit frustrating,' she said, 'Gerry being rationed to a few of the cheapest cigarettes, never going out because we couldn't afford it, but we used to get a lot of fun out of the simpler things, things we never do any more like going for long walks, and I think we were far happier then in a lot of ways even than we are now.

'Surprising,' she said with the wry but fond criticism applied to Durrell by any woman who knew him well. 'Because I really didn't like him when I first met him. I think that comes over in my book, *Beasts in My Bed*, though of course nobody believed it coming from a Durrell – they always forget I'm not a Durrell, so tell the truth. I thought him conceited, arrogant and thoroughly spoilt. And, to crown it all, women liked him, which I always think is bad for any man. And I found him very pushy. He took it as a right to get everything he wanted and frankly no one could refuse him, which I found extremely distasteful, because I certainly could. And did.'

Jacquie spoke with her usual steadiness. She was clear and cool in tone, with a touch of northern accent. Her eyes were a bright brown. Her ears stood puckishly off her head under a thatch of short hair that had bronzed in the sun. The round face ended in a firm chin that suggested both determination and the humour to use it wisely. I had always liked her and her sharpness. I felt at one with her acerbic attitude to Gerry, his excesses, the way he pushed his luck, his tendency to domineer given half a chance, his boyish need to be king of the castle. She was both Durrell's match and more than a match for Durrell. She detested cant. She loved wit. She shared his humour. 'From the beginning,' she said, 'our relationship was rather fraught because of my father's animosity. Not wanting me to get married, he had a habit of

116

seeing liaisons where none existed. Being rather perverse by nature, I always work on the premise that if somebody's accusing you of something, you ought to damn well go and do it, so that at least you're being justly accused. But I must say, looking at it from a rational point of view, my father had no reason to believe that Gerry would be any different in the future from what he seemed to be then – indolent, unambitious, penniless and addicted to drink.

'My father's father died at a relatively early age through alcohol,' Jacquie went on. 'Father himself, given a couple of drinks, turned into a raving lunatic. So, with this built-in abhorrence of drink, anyone who was to his mind a drunkard had to be left severely alone, especially if this chap wanted to get mixed up with his daughter and share his basic evils with her. One has to be fair to my father. Even Gerry admits it was true – he hadn't a bean, he was ill with malaria, which might have afflicted him for years, and he drank whisky a lot, a bit stupidly, almost out of bravado. It was the old alcoholic story of always somehow finding the money if you want a drink, even to the point of pawning the books. He was in the doldrums, convinced that his collecting career was already at an end, and in the state, as Gerry is inclined to be at times, of believing that you can drink any problem away.

'When I first began to take him seriously,' Jacquie said, 'he had a suitcase of unopened bills, which he always carried with him rather like Christian with his burden on his back, and my first task was to go through all these bills in Manchester – the very fact that he allowed me to do it showed the extent of his guilt complex. I went to Liverpool on his behalf to settle accounts over the importation of animals, and by the time we had found homes for his collection there was enough money to pay the debts. Basically Gerry is not dishonest in avoiding his obligations. He just doesn't like to know about them. That's why nowadays, of course, I don't think Gerry signs a cheque from one year's end to the next. I handle all his affairs because he just cannot be bothered. If he has money, he spends it. If he hasn't, he sits down and earns some more. If he can't, too bad. I honestly don't trust him with money because he's a total spendthrift – it's just a

117

commodity, it has no significance, it's merely a means of bringing about things which he wants or needs to accomplish.'

Once alone in charge of his finances, Durrell's habit was to pull from his trouser pocket a muddle of hundred-franc notes to buy toothpaste or tip a taxi driver, never recognising by colour the denomination, or to hold out a handful of foreign change for a shop assistant to pick at. In England, three years after the introduction of decimal currency, he had to ask how to fill out a cheque. There was no trace of affectation; nor was Durrell genu-inely helpless. It was just that nowadays these matters were habitually taken off his hands because they bored or irritated him. Nor did his failure to take any interest in money indicate any deeper irresponsibility. 'Fortunately,' Jacquie went on, 'he has a strong sense of responsibility as regards me, and he also realises that unless he keeps himself fairly solvent he can't keep Jersey going and developing. These two points combine to force him into that abhorrent activity known as "writing".'

She paused, hands lightly clasped on the table, shoulders hunched, brown eyes round and serious, giving an impression she could sit thus for hours without moving, resembling – what, which species? People who met Jacquie were always trying to hit on the animal of whom she most reminded them, a marmoset with those large pointed ears, a squirrel with that bright stare, some form of baby monkey, even a koala. If none of the analogies quite worked, there remained a compulsion in many people to regard her with affection as one animal or another.

Durrell now returned from the telephone and sat down to order a modest lunch of hors d'oeuvres and duck *à l'orange*. His account of their meeting and marriage, if less matter-of-fact than Jacquie's, balanced hers nicely. 'After that first expedition to the Cameroons I went up to sell some animals to Manchester Zoo,' he said, 'so I trotted along to a cheap hotel full of decaying singers from Sadler's Wells and run by a strange man who looked like a Punch caricature of a Jewish financier, short, stout, hook-nosed, black eyes close together, and a large boxer dog at his heels. Soon I was having an affair with one of the opera girls – not particularly attractive but most willing – and I went into the drawing room one day and there sitting by the fire, surrounded by all these girls, was something that looked like a baby robin, and I can remember it to this moment, absolutely vividly, it went boom, just like that, into my face, and I thought feebly, Gosh, she's stunning! And then I went to Guiana. Normally on a trip everyone I knew vanished, but this wretched face kept recurring when I was cleaning out cages or floating down the river in a canoe; all the time it kept flashing into my mind, and I thought, Why should the others be so unmemorable while this one keeps intruding on my brain? So I made a mental note that, should I be able to inveigle her into a corner, I would try and get to know her better – which is exactly what I did.

'I transported the whole collection to Manchester, zoomed straight back to the hotel and took up residence. And proceeded to fight off all her other boyfriends. She says she didn't like me at first. She was certainly very brusque, as only Jacquie can be when she puts her mind to it. However, I persevered and managed to jockey her around in the end. All her other boyfriends were comparatively inexperienced, whereas I was the first semi-adult she had encountered – she was a bit scared and didn't know how to handle me. She knew that I was a man of the world, whatever that means, that I'd actually had "affairs" with "ladies" rather than just a furtive fumble in the clover. So I suppose she found that a bit offputting, fighting a rearguard action in the first place in case what did happen might happen. I then spent six months in Manchester by getting a job in the zoo aquarium to be "near her". Having managed to corner a bit of the market, as it were, I had the additional advantage of being actually in the house. I always believe in carrying the warfare into the enemy's camp. But her dad was terribly anti-me and not without good reason, because I had no talent except for seduction, no money and on the face of it no prospects whatsoever.'

To avoid acrimony or the law, they waited until Jacquie was twenty-one. 'If you're going to marry me,' Durrell said, 'you've got to break your strings to your father, so do it sensibly.' The only sensible way was to elope by rail to the one place where they could get a free roof over their heads, where Mother bought the cake and the wine and Margaret repaid some money Gerry had lent her from his inheritance. 'I think we had about forty pounds left,' Jacquie said, 'when we had settled all the debts, which he proceeded to spend lavishly on me, which I wished he wouldn't.' There remained the bare cost of the licence and the ring and, as Durrell put it, 'we had our honeymoon on the carpet in front of the gas fire, and we lived very happily, very cosily, in that little room, buying everything on no money at all. That's why Jacquie is still so clever at making half-a-crown do the work of ten bob. I've had to fight her over it because when there's no need for such economy it has a very stultifying effect on you. Years after we hit the jackpot she was incapable of spending five shillings without an endless amount of consideration, pacing up and down, argument, biting fingernails. She will spend enormous sums on me, twenty or so presents at Christmas, but nothing on herself. Even her clothes I have to force her into buying. An awful woman in that sense, there's nothing she really wants except the odd record. On Christmas Eve I rack my brains, because if I've secretly bought her something she almost inevitably guesses right, which spoils everything. One year, after due thought and many reminders, she said, "I know, you can buy me a creeper." A creeper? A creeper? It turned out to be a little object on wheels on which you lie down to go under a Land-Rover to service it . . .'

We completed the meal hurriedly. There was a dash to the car. Within an hour, having raced out of Bournemouth after leaving it a bit late, Durrell was seated in the back of a small plane taking off from Hurn. The town passed at a sickly angle beneath, most of the old haunts wiped out by patterns of villas and new streets. The few remaining associations, the things that still mattered to Durrell, could be picked out close to the pier. The Royal Bath where the oysters were luscious and the waiters ready for a joke. The Russell-Cotes Museum, that stuffed, painted, filigreed and magnificently vulgar gift to the town of an Orient-orientated ex-mayor who had brought to Bournemouth the worthlessly priceless collection of a Victorian lifetime. The venerable bookshop of Horace G. Commins, rising dizzily up narrow stairways to room after room packed from floor to ceiling with adventure, discovery, invention, revelation, delight. A very few bars. And then, as the aircraft lifted over the sea, more distant associations. There to the right lay the islands of Poole Harbour, still more or less intact, quiet, given to birds rather than humans. Beyond the little old-fashioned ferry to Sandbanks, the untouched spreads of sand skirting the Isle of Purbeck. The mysterious hills rolling inland to the sudden sight of Corfe Castle on its eminence. The quarries where recently dinosaur footprints had been found. Such old haunts as the archetypal village pub overlooking the sea at Worth Matravers.

It was hard to say how much Durrell was still nourished by these early roots. Usually he denied that England meant much to him. But his denials were often so violent, and his desire to share the pleasure of his youth in an old bookshop or old pub so keen, that it seemed likely he had turned on England only because it had thwarted him: this was the district where he had wanted to establish his own zoo. He had battled with the authorities. He had fought local prejudice. The futile struggle had lasted ages. Only in 1958 did he make his first flight to Jersey, at the suggestion of his publisher Rupert Hart-Davis who had a friend who owned a manor, and within hours of his arrival a zoo was his. Or, as Gerry more bitterly put it, 'At last, bogged down by the consti-pated mentality of local government, frightened off by the apparently endless rules and regulations which every free man in Britain has to suffer, I decided to investigate the possibility of starting my own zoo in the Channel Islands.'

The curved orange beaches of Alderney swung away below us. France was just visible as a hazy coastline, a scribble of pastel elongating the horizon. Brandy in hand, Durrell stared at his favourite France with evident nostalgia, though he had left it only yesterday. In no time the greenhouses of Guernsey were passing beneath. The acres of glazing were interspersed with handkerchiefs of cabbage, and Durrell gazed down at that pattern of islands as if he owned or were on the point of writing about them. Then across calm sea the little plane was descending into the unsteady line of the runway at Jersey Airport. In the setting sun the island looked small, dense, compact, no bigger than a large zoo. Indeed, as Durrell had written of his first visit, 'it seemed like a toy continent, a patchwork of tiny fields, set in a vivid blue sea'. It might have reminded him with a tremor of Corfu; the words suggested it. 'A pleasantly carunculated rocky coastline,' he recalled, 'was broken here and there with smooth stretches of beach, along which the sea creamed in ribbons. As we stepped out on to the tarmac, the air seemed warmer and the sun a little more brilliant. I felt my spirits rising.'

On this occasion two tall men, hiding their shyness behind set smiles, awaited Durrell's arrival as the aircraft taxied to a halt. Lanky and bespectacled, with a brisk humorous face, John Hartley was the Trust Secretary, now in his early thirties, who had started as a boy with reptiles in the Zoo's early days. As Zoological Director the equally lofty Jeremy Mallinson was second-in-command to Durrell. His breezy thatch of almost white-blond hair contrasted sharply with a ruddy face dominated by a nose always described by Durrell as Wellingtonian. He was now grinning lopsidedly as Durrell stepped off the plane. Noises of welcome and homecoming were made. Indeed, despite daily reports by letter or telephone during Durrell's absence, they all started talking at once, reporting, speculating, gossiping. They talked their way into the terminal or rather the charter firm's office, talked through the wait for baggage, talked and joked airily over to the waiting car. It sounded to an outsider like a high-spirited dialogue that had never been interrupted. They talked, more soberly now, all the way across the island to Les Augrès Manor, a successful triumvirate at last reunited on home ground. The mood suggested that zoos were as much fun as agony to run. Or, less generally, that this particular zoo, to the exclusion of private life or leisure, meant everything to all of them.

They talked of certain members of the public who threw into the cages of the great apes such items as lipsticks, razor blades, packets of aspirin. They discussed Durrell's projected visit to Mauritius, island of the dodo, in an effort to save the pink pigeon, of which only two dozen individuals remained in the wild. They spoke of soon constructing a breeding complex for the small

Madagascan predators which were in severe danger from the felling of tropical rainforests and which could only benefit from what Durrell called 'battery farming with a heart'. Mallinson reported progress in his negotiations with Frankfurt Zoo to share a breeding programme for the Jamaican hutia. Nocturnal in habit, no more impressive in appearance than 'a hamster with elephantiasis', this was one of those threatened creatures never taken seriously by the more conventional zoo, which needed giraffes or dromedaries to hold the public. Such zoos cared only in theory for the thousands of less showy species that were dying off. Meanwhile Hartley said that a further grant had been awarded to the two researchers who were caged for hours at a time with the baby gorillas. They were introducing them to toys, and making a study of their psychological advance in comparison with human infants. A tragopan pheasant, dying inexplicably, had gone for autopsy. The offspring of the bare-faced ibis, which in the natural state had been reduced to nesting on one cliff above a Turkish village, were doing well. In one year Jersey had almost doubled the world population.

We halted amid the spaces of an empty car park. The day's visitors had departed. Still talking, the party walked briskly down the slope, past a perky mynah that croaked what seemed like human greetings. A family of peccaries snorted in their mud. The young orang-utans were now shut away for the night. The manor house stood in a hollow square, a rock-solid edifice that commanded the thirty-two undulating acres that belonged to the Zoo. It looked like a rough-hewn sketch, in almost coralline granite, of a country residence, English, eighteenth-century, made of a stone containing, in Durrell's words, 'a million autumn tints where the sun touches it'. Flocks of black-and-white pigeons circled the rooflines over which flew a tangled Union Jack. With a raised eyebrow Mallinson complained how tattered it looked. 'This is not the British Empire, Jeremy,' Durrell said. 'We have a future here, you know.'

He crunched over the gravel to the front door. The pigeons settled for a moment. The ground floor, a muddle of offices that had finished business for the day, had the fusty air of a school on holiday. Acres of desks stood in once social rooms adapted to office life. Notice boards had announcements and jokes pinned to them. A gloomy corridor, piled with annual reports, wine cartons, milk bottles, led to a flight of stairs. Another door opened into the lighter, softer and vividly coloured privacy of the Durrell flat, the single perk of his honorary office as Director of the Trust. The expansive kitchen was painted a flashy orange. It was ideally serviced and equipped. Serious cooking, after all, was Durrell's favourite pastime.

In the distance lions grumbled. There was an abrupt and faintly disturbing shriek, as if of pain. But the noise died quickly with the dark, muffled by bourgeois comfort. The chairs were deep. Vibrantly the colours of wall, carpet and upholstery clashed. Glass-topped tables sharply reflected strong light from shaded lamps. Teddy bears and woolly squirrels squatted on the edges of shelves, where books in their gaudy jackets were stuffed from floor to ceiling.

Durrell had reached home, in so far as he had one, and happily enough, as far as happiness was a concept that interested him. And he had much to do.

Thursday

THE ZOO woke early. No cock crew at dawn. Instead a muffled vocal combo of jungle, pampas and steppe, with a hint of nightmare, grunted, screamed and hooted out of the sunrise. For the human beings it was like springing awake to the demands of a huge, exotic family who would grow obstreperous if not tended at once. Durrell was always complaining about the lack of suitable accommodation for his men and women; the shortage of funds meant that they fared worse than most of the animals. But certain members of the staff lived on the spot in a cottage attached to the manor house, and they were already at work well before eight o'clock. A tractor started up. Someone was distibuting loads of straw. Others pulled hoses from one complex of cages to another to swill them out. They were chopping fruit, nuts, vegetables, meat (of a quality superior to their own hasty breakfasts) into dozens of trays, bowls and saucers, even tobacco tins and old pieces of crockery from charitable sources. Devotion to the cause – nobody could or would work here without sharing Durrell's insane dedication – was disguised at this early hour by a sleepy veil of humour. The cages of the great apes were occupied by bare-chested youths in jeans, removing excrement and orange peel from the sloping floors with jets of water, singing or talking to themselves as they worked.

A few birds screeched, bulleting with a flash of tropical colour into the ample foliage. The cages were designed both to reproduce their habitats and give them a precise sense of their territory. Meanwhile upstairs in a carpeted bathroom a human figure stood upside down on his head, stomach sagging towards his chin, eyes glaring upwards from floor level. Durrell was doing his yoga. He too had started early.

On an upper landing in the flat were stacked dozens of copies of Durrell's books in many languages. A best seller everywhere, he had carried his one-man crusade to the world. 'Durrell almost convinces you that life is nothing unless you own your

127

own zoo,' said an American blurb for *Menagerie Manor*, Durrell's 1964 account of the Zoo's daily routine. These words were to seem more than almost convincing after a day spent within the spacious, busy, raucous but strangely placid confines of Les Augrès Manor.

Shortly after finishing his yoga Durrell was drinking coffee at the long wooden table in the kitchen. He still appeared to be suspended in meditation. He looked forward to the day. But anxiety counteracted with nervous self-mockery his fear that today would prove, as usual in Jersey, a series of irritations, if not a disaster. 'I'm a creature of habit,' he said, looking remarkably unlike one. 'I get disrupted very easily. If I know someone's coming to see me midmorning, I can't settle to anything until that person's out of my hair. Thus a not very important encounter with, say, Jeremy Mallinson actually consumes four hours – entirely my fault for being unable to do anything constructive with those hours. I dabble, dictate, doodle, chat, feel fussed – which is ridiculous. But I have to follow these patterns. With Jeremy, for example, I adjust to his timetable, simply because his routine in this place is the animals, all the animals, and nothing but the animals.

'What I do to obviate it is get up early – five o'clock, yoga, a hot bath, lashings of tea – so that from six until the post arrives I can proceed with the big yawn of putting words on paper with nothing to distract me. But, oh, it's so much more fun to go and watch the gorillas than pen another glowing paragraph!' He paused. 'What infuriates me now is that I spend more time on administration than I do on the animals and that isn't what I created this place for. I don't seem to know them any more.'

To make matters worse, the problem of being recognised by visitors also cut Durrell off from his animals. He was dependent on visitors for the realisation of his work – their daily patronage accounted for a good percentage of the Zoo's income – so he never minded when people bore down on him or asked for an autograph. He was shy, indeed humble enough to feel it ridiculous that he should be fêted for so little. But fans did prevent his concentration on his cages. By the time he finished his dawn writing, early enough to steal a march on the pressures of admin, the Zoo was for much of the year already crowded with trippers, most of whom had read his books. By the time they departed it was dark or the animals had been shut away for the night. It was particularly sad since he had formed intimacies with many of the creatures when capturing and nourishing them in their countries of origin. In a sense Durrell was the most miserable captive in his own Zoo. Every time he stepped into the courtyard during the hours of opening, perhaps to show round a guest, he had to steel himself to the task.

'I can't really explain why I was attracted to animals in the first place,' he said, sipping his coffee, 'but it's immensely strong, always has been – it blots out everything else. I suppose it's like being born with three legs. You're so used to it that you never give it a thought. I just know that I seem to have it much more intensely than the average person – though I believe everyone has it (or should have it) if you dig deep enough. It's probably the way a real author would feel about his writing. But this is not creative, it's a matter of observation. I'm a translator, if you like: watching animals and translating them on to paper in the hope of helping another person who is not so familiar with the animal. My intense interest is so enormous inside me that I imagine it to be the sort of feeling a painter has, simply that he must be a painter and nothing else, even if he fails – whereas he might succeed better at something different. I don't want to be a good writer. I'm an animal man.'

In the background a lion roared, a rare lion from India. 'Mammals come first for me,' Durrell said. 'I love birds for their beauty, but only with the more intelligent echelons of the crow family can you have the degree of rapport you get even with a squirrel, which is much higher up the mental ladder. I've always been bowled over by the sheer diversity of small mammals – a marmoset, a lemur – the way they adapt themselves, so small, so fragile, to so many niches in life you would think quite untenable. It's rather like a hummingbird. You ask, you ask open-mouthed, How does this creature exist for more than thirty seconds? It's like gossamer, you know, it's like dew. And yet they've fought their battles to exist, they're tough little things, they live in their own world – and they've got all their rules and regulations set down, they live by the book. Rather like trade-unionists actually, but a bit more charming.'

With a grin Durrell set down his cup. As always when his mind began running along favourite lines, his face grew tense with enthusiasm. 'The great apes fascinate me most of all,' he said. 'It's like being lucky enough to be plonked down in a cave full of Neanderthal men. You have the extraordinary privilege of watching the patterns of their behaviour, the shifts of their thought. They're a separate branch of the primates, of course, nothing to do with man. But I would imagine that man in his very early stages had similar mental processes to gorillas and orangs in particular. So while you watch them – it's very solemn for us, with our minor advantage of our thoughts being conscious – you're quietly thinking to yourself that you might well be watching primitive man.

'However, it's difficult, very difficult, to interpret their minds and feelings with any accuracy. Chimps are much less private. That's why they have proved more useful in experiments. They're the extroverts, the professional actors of the ape world, crying out for an audience the whole time. Show a chimp how to wear a hat and in thirty seconds he's Shakespeare, whereas the orangs and gorillas are the thinkers, they sit down and ponder a problem. And the fact that the primitive structure of their lives so clearly echoes what we know of early man makes you wonder – if man was removed from the map (not a bad idea), if the world went on ticking without pollution or man's interference for another trillion million years – makes you wonder what these great apes would ultimately turn into. Would you come back and find a gorilla in charge of West Africa, an orang-utan sitting in Djakarta? If only they were allowed to evolve!'

Durrell paused. His eyes, which brooked no contradiction even in a realm of fantasy, were hard and bright. 'But the point is they won't evolve into anything now. The total world population of orang-utans is down to (let's be generous about it) ten thousand. With human births rising at the present rate, all the forests that house them will be chopped down and turned into farmland that will soon be a desert, all their trees will be processed for one reason or another, generally to print lies on, so where will your orang-utans live? If they exist at all, they'll exist in places like this.'

Durrell heaved a sigh, rose abruptly to his feet, and started to assemble on the cutting board a battery of cubes, herbs, tins and vegetables for one of his square-meal lunchtime soups. He scraped and chopped the carrots, sliced the onions. He was using the silence to let the point sink in: that there were creatures which, given a chance in an unspoilt world, might turn into animals much superior to man. A large saucepan of water was placed on a low heat. The silence continued. He had preached the animal sermon. Now it was to be the human turn. He sprinkled the warming liquid with thyme, coriander seeds, black peppercorns, chicken concentrate brought back from French supermarkets. 'And if they don't continue to exist,' he went on, 'we won't. We won't. Because human beings are wild animals. We all make a whole, we're all parts of a jigsaw. We men have chosen to pull ourselves out of the puzzle in order to pretend that we're God, we're strutting and fretting on our little stage for the moment, but we're already running into trouble, trouble that can't be measured with our little brains. You have only to open one eye to see that what any intelligent conservationist has been preaching for years (and years and years and years and years) is now coming to pass – and everyone said they were alarmist.

'You cannot drag yourself out of nature like this. It's just as important to preserve an ostrich as a reporter in Fleet Street, in fact on some occasions probably more so.' Durrell poured diced vegetables deftly into the saucepan. 'But we have this awful insular habit of talking about the animal kingdom patronisingly as though it were an inferior form of life on a very remote planet – but we *are* in the animal kingdom, we're a bloody mammal. And the whole educational system supports this appalling divorce, as do politics, social custom, even science – it's all geared to this frightening idea of man being God, instead of the worldwide

Frankenstein he has really become. I don't think we should all start wearing woad and living in caves, but at least for God's sake let's respect and honour our world and understand how it functions and how you are supposed to function within it, so that you don't constantly kick it in the teeth with every movement you make, which is what we're all doing.'

Hesitating to interrupt, John Hartley put his head round the kitchen door and stared at Durrell with raised eyebrows and an interrogative smile. Durrell nodded, stirred the soup once and tramped downstairs. Twitters not unlike an aviary emerged from the general office. Frequent butts of his humour, the female staff had arrived. They were taking covers off typewriters and stowing handbags, mingling in gossip the accents of Scotland, Jersey, the West Indies. Durrell stood for a moment at the sliding door. The girls bustled, giggled, welcomed him home, wished him good morning. He continued to stare balefully as if maturing a wisecrack, thought better of it, and turned into his own office, where Jeremy Mallinson and John Mallet, Curator of Birds, were gathering for a conference about the trip to Mauritius.

The room, lined with shelves containing a lifetime's collection of books on natural history, was furnished with a small desk for a secretary, a larger desk for Durrell, an armchair for a plump cat, a filing cabinet for cuttings, projects, reviews, scripts and outlines, and a flashily suburban drinks cabinet to entertain visitors. French doors opened on to the courtyard, lace-curtained against the curiosity of the outdoor overspill from the Zoo's café which, to Durrell's irritation, occupied premises in the manor house. The room's mood was stuffy, disordered, raffishly academic, the haunt of a chaotic don. Mallet had wellington boots on, smeared with birdlime. Mallinson was wearing a school or club tie. Durrell sat at the desk with the half-assumed look of a worried tycoon, a reluctant figurehead. John Hartley closed the door against interruption.

Durrell had total trust in his henchmen. He wanted to elicit their views, either to confirm that they coincided with his own or to be taught something he did not already know. He had trained these men from boys who bred hamsters into serious

132

zoologists; or, more accurately, he had let them under a light rein conduct their own training. He would permit no opposition to his own clear and often inspired ideas for running the Zoo, which was his conception, his creation, his joy. Never pretending to be a judge of character, he relied on intuition for his choices. He would never employ for more than a few months anyone not on his wavelength, any more than he could tolerate someone who showed neither independence of mind nor initiative in action. He asked a lot.

'I'm not a yes-man,' Mallinson whispered, 'but I *am* a disciple.' Disciples were born from acts of conversion; they then interpreted the gospel in their own manner, often eccentric, always consistent. The three men present in the room, after talking over methods of trapping fruit bats, started to plan the strategy for the forthcoming expedition to Mauritius. By nature cautious, Mallinson took the view that it was preferable to concentrate on six different species in the long term than to bring home too many, however much threatened, and thus be forced to expand in too haphazard a fashion. Hartley's temperament convinced him that, however many species were found in need and individuals caught, short-term aid would always magically be found to house and feed them. By the time the pigeons, kestrels, parakeets, were out of quarantine, he argued, someone would have been per-suaded by his well-known fund-raising techniques to provide another range of aviaries. Mallet remained silent except when jokes were in order or birds on the agenda. By his own admission he was less of a fanatical conservationist than his two colleagues. He had merely loved birds since his Hampshire childhood and preferred to help them survive rather than see them go under, an approach more aesthetic than philosophical. As they talked, their views united into a blend that best expressed Durrell's own mixture of caution, rashness, sense and vision, and the meeting drew to an end.

Durrell stretched. 'You see, we appear not to have decided anything for certain except that everyone present was in total accord,' he said. 'I trust them as much for their faults as their virtues. I place absolute faith in them. If I died tomorrow there

133

would be no fear of collapse through a lack of directional impetus. It's all spread. They might regret the loss of a fund-raiser, even of a father figure, but they would know just which way to go.'

Jacquie meanwhile was driving into St Helier, which Durrell saw as a staid English market town trying to imitate a tropical paradise. Palms grew in certain parades. The place conveyed the lazy mood of long lunches funded by secretive accounts lodged in the many banks. Having detected a minor fault in one of the cars, Jacquie spent ten minutes affably browbeating the garage man, who listened in some awe to the extent of her technical knowledge. She drove on into town. Terraces of petite Victorian villas, neatly kept by the retired and suggesting a leisured spa, lined the slopes curving down to the old-fashioned centre, a capital city in miniature, where merchant banks nudged duty-free shops in alleys that sustained an airy seaside gaiety, just a touch of the continent. 'Getting away from here as often as he can,' Jacquie said, 'is the only way Gerry is going to survive in the future.' She coasted past a covered market packed with flowers for export, expensive cut-price jewellery, local crab. 'He feels France is within easy reach, should he be needed here. Now he has become utterly confident in Jeremy and John Hartley not to bitch things up in his absence, perhaps he can start turning his attention from day-to-day problems and concentrate on getting pleasure out of the animals. I have a very simple philosophy: if you get no joy out of what you're doing, go and do something else. He doesn't believe me, because he knows it's true.'

She raced the car up to the second floor of a multistorey car park, yet another approved convenience of the Durrell world along with supermarkets or motels, and jogged back down to street level where she set off at a jaunty pace for the market to buy the foods Durrell liked. Despite an apparent detachment she knew exactly how best to please him and was determined to do so. She bantered with the fishmonger while taking her time to select Gerry's lobster (and a trout for herself). She exchanged some jolly crossfire with the lady who ran the fruit stall, as if shopping for men's appetites were a genial conspiracy that only women understood. At every port of call she left her parcels behind to

134

be collected at the last minute and at every point, not exactly suppressing her own tastes and desires but adjusting them to those of her lord and master, she thought and spoke of what Gerry might or might not like, whether a wine would appeal to his palate or whether a peach was ripe enough. 'We're totally different in outlook and temperament,' she said laconically, 'but I think it's the way we were brought up. If you really analyse it, there's nowhere we have common ground except on animals and the fact that we like to travel – we even like to travel to different places. For example, we both like music, but Gerry likes what I call piddle melodies and I care for the round, luscious music of opera and the so-called romantics, and I adore modern pop, which Gerry hates. We both need music as a stimulus, but while I go for total contrast Gerry tends to get in the rut of a vogue. It's all Vivaldi and I get driven absolutely potty, or it's all Greek or South American, it's never a variety.'

Her shopping finished, Jacquie now ambled speculatively past the windows of one or two department stores, looking for clothes, never finding just what she wanted. She paused, stared beadily for a moment at a skirt she quite liked, then dismissed it, turning on her heel. 'I'm more catholic in my tastes,' she said. 'Though you get the impression with Gerry of an ebullient rebel, he's one of the most square people I know, very rigid in conformity, conservative in his views and outlook.'

The popular image of Durrell taking on the establishment single-handed was plainly not shared by his wife. She thought he could be understood only if you forgot the riotous surface of his upbringing and dug out the family roots which were unmistakably middle-class. The muddle of his creative childhood had concealed but not removed much that was reactionary in Durrell – inherited perhaps from his mother, in whom only a sense of humour had mitigated a faith in strict puritanical standards. Such standards, based on a style of life gone for ever, had passed into the Durrell children. Neither an open-air existence free of restrictive conventions nor cavorting in Bohemia had altered those standards. Success – especially perhaps with Larry, to whom material prosperity came late – had quickened the flow of hereditary conservatism in both the brothers. Naturally they wanted to keep what they had worked for: which simple view pushed them towards the outer limits of the right wing. Larry had recently been known to utter sentiments so fascist that even Gerry, no novice when devising speedy methods for political reform, had blenched. Both men were intolerant, but their tendency, at least in talk, was to label as human stupidity what was better understood as human weakness. But at a different level their work equally and entirely looked forward. Their books denied perpetuation of the class war, were in favour of women, believed in progress, whereas their late-night talk denounced the current order; their world, such as it was, was going to the dogs. They were horribly outspoken, especially when together. They spat rage. Their faces contorted with derision. It seemed the poisons of a 'proper' upbringing and a 'selfish' success only seeped out when they were watching television or drunk or alone or with friends who would indulge them. Both suffered from, or rather relished, this quite nasty tendency.

'I think Gerry's real interest in politics,' Jacquie said, 'springs from his concern for man the animal — he is deeply involved in any aspect of the evolution of man as an animal, and that means politics too. We always call stupid anything that we can't understand, don't we? Yet he can understand, if only he tries. He just despairs of us all. He's incapable of realising, for instance, that quote the working man unquote, whatever that might mean today, is locked in this bitter struggle with what he considers his lifelong enemy. Gerry can't see, having never been involved, never having been compelled to work on that level, that they are constantly fighting for their existence and are therefore resentful, because they too cannot understand, of anyone who is not a manual labourer. When Gerry meets a human being, he loves that human being. But in general he is infuriated by their asking for another pound a week. Yet, you see, it's the old pecking order, it's with us for ever, that's what we are like — Gerry's a biologist, he should realise that they want the strata, just as an animal wants his own territory at the expense of others, and the only way they can divide themselves within their own area is by having a pound a week more than the next man. The trouble is that he gets society thrown at him morning, noon and night on television, and he will insist on watching. No wonder he gets irritable.'

Jacquie retrieved the Mercedes from the multistorey. Briskly she collected her shopping from the various points where she had stowed it. We drove back to the Zoo. I marvelled again at the patches of wildness that gaped now and again within such a compact island. Then a number of bungalows punctuated a lot of potato fields. You kept aching for, and getting, a glimpse of the sea, even of France on a good day. The island was always changeable. Sunny weather might end within seconds in pouring rain.

As a sudden breeze blew up, Jacquie shuddered. She appreciated Jersey as the place that had realised for Durrell his prime ambition, but loathed it for what she thought it did to his inner temper. 'That is what I never understood about our return visits to Corfu,' she said, 'because he used to say, "Oh, Jesus, why don't people let me alone?" and yet he knew in advance that he would never find any privacy. It's a weird attitude: if I go near the fire I'll get burned, but I'm going near the fire. It's almost like the old sirens whose voices he can't resist, though he knows bloody well he's going to be unhappy while he's there. And also here, or wherever his emotions are involved. When problems arise in a paradise of your own making, it's awfully difficult for someone to cope, if he's been brought up to believe that everything is marvellous and sweet and sugary. Gerry always thinks an awful lot of it is due to his failure, and being a Capricorn he can't bear to fail, and incapable of accepting failure he moves relentlessly towards a nervous breakdown.'

Which is what happened in 1968. But was it a breakdown? However you regarded it, the root causes were embedded in the past when, for Durrell, life set its standard of being perfect or instantly perfectible. He had believed that once he established his zoo, the basic problem of how to live life with panache would be solved with ease. But at that point, for financial reasons, to gather material for a book, Durrell had been forced at the wrong moment to travel in Argentina, which meant trusting other people to look after things at home. Their task in Jersey was to make the arrangements for a collection of animals on the principles expounded by Durrell. Those in charge during his absence, being human, 'human' in Durrell's lexicon being a synonym for 'idiotic', chose to ignore his instructions in favour of a zoo that suited their own ideas of caging and diet. He bottled up his resentment. He resisted the notion of autocratic takeover because at no time – then or ever, said his friends – did he wish to be tied down. This point he proved almost at once by undertaking an expedition to Australia.

The troubles went back to the autumn of 1962. Under well-meant mismanagement, the Zoo not only tottered to the verge of bankruptcy, it also nearly destroyed the spirit of the people working for it. Both Mallet and Mallinson, distressed by the betrayal of Durrell's original vision of the place, were on the point of leaving. Old Mrs Durrell, ignored and powerless in the manor-house flat, was worried to the point of collapse by the chaos. The accounts showed a debt of £14,000, quite apart from the personal loan of £20,000 which Durrell had persuaded his publisher to put into the Zoo.

The Durrells returned at speed from the Far East to face five days of reckoning. Those days probably took more out of him than he would ever care to admit. Yet almost at once, within weeks, the right people, all island people, arose like genies out of the rubbed lamp, as Gerry put it. Catha Weller responded to a local ad asking for someone to keep the books. For several years her tough common sense kept the establishment on its feet and the Durrells much amused. A former head of Shell in Latin America, James Platt, suggested and launched the simple idea of an all-island appeal, with full support from the *Jersey Evening Post*. This mobilised the concern of a lot of rich names. Island politics joined hands with island society. The well-lined pockets of those always accused of evading tax elsewhere in their big walled mansions were 'generous to a fault', whatever that might mean. But those of lesser income who really owned and lived in Jersey were no less forthcoming. A sum was raised slightly in excess of the debt. Given that such an unlikely concept as 'Durrell luck' existed, it came to his aid in those nervous weeks when the one and only thing he had always wanted, as infant, boy and man, seemed about to founder.

Liquidity assured, Durrell at once relaxed. In moments of stress his habit was always to take the problem hard, then pretend to himself that no problem had existed. This was one way of sidestepping a nervous breakdown in the nick of time. All it did was save it up. It would have been healthier if he had admitted anxiety, and recovered from grief at leisure. He had this ability – didn't we all? – to suppress what ought to be admitted, discussed, even treated. His mother's death in 1964, as unexpected as one of her remarks, though at a great age, hit him hard. 'It was a great emotional wrench for Gerry,' Jacquie said, 'and again, instead of collapsing at that moment, as everyone said he would, including our doctor in Bournemouth, he didn't. In other words, he behaved precisely as Larry did over his wife's death. He wrapped it up.'

The deeper Durrell sank into gloom, the more easily he got away with it. Mutely this eloquent man was crying out for help, his very silence a suicide attempt. His colleagues – all disciples, after all – merely thought that they were somehow at fault. He grew intransigent, contradictory, despite the fact that by now the Zoo was solvent and developing. To all intents he appeared to be functioning well. Books got written, decisions made, money earned, animals bred. 'Yet he was so dreadful to live with,' Jacquie said. 'Nothing you could do was right. "Oh, look, it's clearing up," you'd say. And he'd reply, "No, it bloody well isn't." It reached a point where people were really frightened of him. I was the only one who would stand up to him and do battle. And eventually in 1968, in Corfu, the tremendous backlog of pent-up emotions broke through the bad temper and profound depression and heavy drinking and not talking to anyone for hours, and the rigid control he had maintained over himself for so long cracked.

'And at last he told me what was wrong with him. He suffered these terribly black moods, he felt suicidal. I said, "Why on earth didn't you tell me? I had no idea it had got to this pitch."

I personally thought that what he needed was a total break from the Zoo for perhaps two years, and I was on the point of suggesting this when he reached crisis point, and we hastened back to Jersey. The doctor, when I told him, was all for leaving it until Gerry raised the subject, and I said, "If you wait that long, mate, he'll probably hang himself." And that's when we got him for three weeks to a psychiatrist in a so-called nursing home, a hideous place more designed to give you a nervous breakdown than to cure it, where this bright spark of a doctor told him he was an alcoholic, which I violently disagreed with: an alcoholic is someone who can't live without it, and Gerry can and does. He also told him what anyone with half an eye had been observing for years, that he was a bit overwrought by events at the Zoo. Here was obviously a man only used to stock situations and incapable of dealing with someone like Gerry.

'Anyway, Gerry then decided to treat himself. I think he got over it by virtually bringing himself to death's door, though probably the drugs – he took them for a year – did help a bit in turning him off, relaxing the mental pressure, giving him time to stand back and assess his own capacity for self-healing. And then of course he took up yoga, which has been a great help to him.'

Jacquie swung the Mercedes into the Zoo car park, where a string of coaches waited to return visitors to St Helier hotels. We drove past the high colobus cages where these long-tailed monkeys, which Durrell had brought back from Sierra Leone, sat tense and hunched on the branches in family conference, and halted in the courtyard of the manor house. Upstairs Durrell was putting the finishing touches to his rich soup, sipping, smacking his lips, nodding. As always he seemed glad, even relieved, to see Jacquie back on home ground. Jeremy Mallinson, fingering his tie, stood in the doorway with an expression of bemused formality. To this day he never called his friend and mentor anything but Mr D or, at his most relaxed, Gerald Durrell. Now he was reporting in public-school tones about the results of an autopsy on a peahen.

The Durrells sat down to lunch. They consumed it in silence. They were listening to *The World at One*'s daily summary

140

of disaster in the world at large. Now and then there was a gruff interjection from Durrell, mouth full of brown bread and butter. Mallinson backed tactfully away and returned to his own disordered office, where the Zoo's records were kept in overstuffed filing cabinets. He sat down, the peahen still on his mind, and consulted the daily diary in which any event in the life of every animal, from mild diarrhoea to sudden death, was noted down by the member of staff responsible.

Jeremy's day, after a cooked breakfast, had begun at eight o'clock with a tour of inspection on foot. Tall and slim, military in bearing and ruddy of face, he strode with an air of eager authority from the apes to the finches to the marmosets, bending but not pausing to pick up sweet papers or soft-drink cans dropped by a careless public, a staunch colonial officer keeping an empire clean. With his taste for sailing and games, driving sports cars fast, playing Noël Coward's smoochy swing on the piano, it was no surprise that Mallinson's first choices of career, before discovering his flair for animals, had been the family wine business, the Hong Kong police or tea-planting in Assam. He was cut out by fate for the upkeep of a system that happily declined just in time for him to find a true vocation.

In 1959, just before committing his future to any of those careers, Mallinson happened to be in Jersey when Durrell was starting the Zoo. He took a temporary job, mucking out cages and slicing bananas, to give him time to decide what to do about his life. At once there was no choice. Here he still was. He was conscious of a debt of gratitude to Durrell (which he said would make Durrell snortingly mutter 'sycophant') on two counts. First Mr D had formed his formative years. Then Gerald Durrell had shown the necessary trust to leave him as Zoological Director not in sole charge but to his own devices. 'I'm the one who has most benefited from what he has created here,' Mallinson said, 'in the sense that I am daily concerned with what he would love to be doing himself, but can't, because of his other activities. I'm a parasite on his creation.' On the other hand, the evidence suggested that Durrell himself regarded the finding and keeping of

Jeremy as one of his typical strokes of luck, without which the Zoo would never have found its feet or kept them.

Mallinson's morning had been busy. His mornings always were. On his tour of inspection he had noticed the cheetah sign blown over by the wind, so carried it for repair to the workshop. He examined with care the parrots recently arrived from St Lucia, rare birds in the awkward process of adjusting to a new climate, if not to the idea (were they aware of it?) of captivity. He made tracks for the mammal house, feeling in his pocket for the supply of ginger biscuits intended for his own midmorning break. He fed half a biscuit to each of the mature gorillas. Now and then Jambo managed to copulate between the bars with his female neighbour. But because the male was separated from his two mates, they all needed as much social contact as possible, a continuity of response from an individual who had known them for years.

Finding a tangle of cut branches outside the door of the lemur house, Mallinson reprimanded a member of staff by pointing out that the Zoo was not designed as a commando course for the public. To people who visited perhaps only once a year the entire place must look always neat and well tended. In a subtle way the condition of the grounds should reflect the health and cleanliness of the animals. Dropping into the workshop behind the mandrills, who were gnashing their teeth at nothing and flashing their purple posteriors at one another, Mallinson sketched out for the odd-job carpenter the type of caging he required for the crickets that were now to be bred in quantity, an urgent task because young birds, recently hatched, relied on insect life and were greedy.

Every task was urgent. The pleasure in performing it was clearly sharpened by the fact that in all matters relating to the animals, time was at a premium. You could neither neglect nor delay. It was easy to see how much Durrell was missing these days by functioning at a remove from the surprises of daily routine.

Back in the office, where time for deskwork was unpredictable and usually short, Mallinson between interruptions coped with the morning mail. A wallaby had just been found dead, setting in motion arrangements hardly less complicated than those following a human death. The skull was to be preserved for the Zoo's educational programme, the vet was telephoned to be offered heart, liver and other organs he might want for research, a fecal examination was put in train. All this, at the cost of half a dozen brisk conversations, was entered on the animal's cards. Next, a stillborn lemur foetus was sent to the Wellcome Foundation in London, while urine samples taken weekly from the gorillas were prepared for despatch to labs in Edinburgh. 'Nothing is wasted,' Mallinson said. 'Anything that dies is given to whoever has a use for it.'

On his bike he visited the site of the new reptile house. He suggested the removal of a heap of woodwork discarded in the reconstruction of the bird kitchen. He passed on to the unfinished marmoset complex, where he corrected one of the petty errors that builders invariably made when left alone for an hour. He measured out the quantity of wire needed for caging the marmosets. He ordered several cases of beer for that evening's staff showing of a film on rabies. He alerted the local press to the presence at the Zoo of eighteen members of the Conservation Corps. These volunteers, devoting their holidays to the environment, were removing some old trees overhanging the lake and endangering a rare species of waterfowl.

I was still watching, tracking, listening, slightly exhausted by now. Nobody got bored who was as busy as Mallinson. Routine as various as this produced only excitement. It gave you a sense of instant progress. By now he was ordering the wire he had measured for the marmosets. By phone he fixed a fencer to come on a morning three weeks hence, which would just about coincide with the builders finishing the cages. At the same time he booked the same man's next free afternoon to unravel the tattered flag which gave such a bad impression to visitors gazing up at the manor house.

Someone from the Conservation Corps dropped in to ask for a forty-foot hawser to prevent a felled tree from crashing into the pond. Mallinson tried one firm without result, then another. At length he ran to earth a bulldozer contractor who would deliver the hawser within the hour. Here an unlikely moment of calm enabled him to check the mock-up of the new Zoo Guide. This contained six months of photographs of different species, linking them in simply understood terms to the overall purposes of the Zoo. He dictated a memo on the subjects to be covered in the accompanying text. He skimmed through the wording for a display board that informed the public about some recently planted trees known as arboreal fossils because living specimens had only just come to light. He asked the maintenance man to construct a wooden backing for this notice and commissioned a waterproofing firm to glaze the front. He listened while a member of the mammal staff, that day returned from a sabbatical month at Basel Zoo, described a family tree clarifying the fatherhood of Jambo the gorilla. Someone else reported an abscess on the upper jaw of one of the peccaries. The card was checked, the vet rung. This peccary had been treated earlier, having fared badly in a quarrel with a mate.

The office for once was suddenly empty. There was a brief respite punctuated only by a secretary tapping out letters in the background. 'I have to programme myself because basically I'm lazy,' Mallinson said. 'And at the moment I'm having to squeeze in four hours on a book every day from eight to midnight. In the type of work we do here one puts the whole self in – or forgets about it. There are no two ways. All the time I've recently used for my own writing I now want to allocate for six months to really learning the piano . . .' Someone knocked, the phone rang. 'I'm infuriated by complacency and bad manners and I have no sympathy for stupidity,' he said, as if echoing Durrell. 'Otherwise I'm doing exactly what I want, and I know how lucky I am to be doing it, especially as I love Jersey and, to prove it, I'm married to an island lady.' Mallinson gave a short, hoarse burst of laughter, then looked briskly at his watch. It was lunchtime. He stood up, left the office at speed, drove fast to the local pub, consumed a helping of cottage pie and baked beans, drank a pint of Guinness, and returned eagerly to an equally unpredictable and fast-paced afternoon.

It was usual for John Hartley to show round the Zoo any

special visitor – member of the Trust, sympathiser from foreign parts, potential supporter. His manner was easy, humorous and informative; cheques came his way. When visitors more distinguished – fellow zoologists, delegations from afar, the obviously rich – arrived on the scene by appointment, Durrell himself was induced to emerge, at first shyly, unwilling to face strangers, and then with growing confidence and pleasure in showing off his obsession. Such a party of well-wishers appeared this afternoon. On shaking hands Durrell as usual took an immediate liking to them. They began the grand tour. It was almost as though Durrell himself, in describing it for others, were seeing the Zoo for the first time himself.

He certainly never behaved as if he owned the Zoo. But he did want to share with others the privilege of his own daily pleasure in the animals. He walked with his party, setting a brisk and nervous pace, to the marmoset complex where he proudly stood aside for Mallinson, who had designed it. The point was, and Mallinson made it well, that this environment suited marmosets just as well as the wild state and possibly, in such ways as protection from predators and possession of their own unthreatened territory, even better. With care Durrell listened, as if hearing news, to this precise account of how much thought had gone into supplying these needs, creating adjacent areas in which the marmosets could ogle and be aggressive to one another without injury, providing them with inner retreats where they could rest, reproduce, sleep, away from the public view.

'This,' Durrell then said, taking the stage as lecturer, if not evangelist, 'is a prototype of exactly what I want to do for all the small mammals in groups that we have. When I think of the awful dark damp dingy cages in which marmosets used to be kept, I now realise we have a chance of giving them freedom.' It seemed an original concept, actually to liberate an animal by incarcerating it. But, as Durrell pointed out, an animal's attitude to imprisonment could be judged only by the quality of its fur, the brilliance of its glance, its general vigour, and most of all its active sexuality. 'You have to feel secure to want offspring,' he added.

The party moved down the slope past a slim lake where

waterfowl were active. For the visitor they were identified on ceramic plaques made by a local pottery, each of which was donated by a Trust member responding to newsletter appeals. On the more explanatory notice boards the birds were delicately drawn and coloured by a member of staff. 'Mustn't encourage him too much or he'll take up as a bird artist and leave us,' Durrell said. 'Ambition has to be strangled. But we're jolly lucky.' Discussing the cost of these notice boards and how difficult it sometimes was to justify the expense, Durrell walked down to the outdoor cages of the great apes, one of the climaxes of the grand tour. Here as usual a handful of visitors was gathered, giggling, imitating, identifying. As usual, too, the apes were behaving with more sense and dignity than the people. Leaning on the rail, Durrell fixed an eye on Gamba, a shaggy and beady-eyed orang-utan, and began to croon, 'Come on, Gamba, walk around, show us how beautiful you are.' Gamba did not move. But the tension was obvious, an almost hostile but direct response that at once commanded silence in the small crowd. The man stared at the animal; the animal stared at the man. The man did not assume himself to be superior. Each was trying, by guesswork, intuition, a touch of the aggressive, to bridge the gap supposed to yawn between two equal creatures who shared no common language other than that of being alive. The gap appeared to lessen in the silence. Then Durrell drew back from the encounter. 'He's a marvellous beast,' he said. 'Touch wood he should go on for another twenty or thirty years. His mate over there is an absolute bitch as a personality.'

'Excuse me,' said someone in the crowd, 'is that the mother?'

'Yes, the little one,' Durrell replied courteously. 'She's already had a baby, you'll see it when you go round the other side of the manor house – and the gorillas have both had babies and now they're pregnant again. We're sex-mad in this zoo.'

Everyone laughed. Everyone shifted a bit closer, looking at him with a blend of awe and sympathy, as if expecting him to let them into secrets as well as crack jokes. Their interest fed his enthusiasm. Almost tripping over his words, he got down to such

146

practical detail as the widely spaced bars of these mammal cages. 'They're super for the babies, they climb up and down like mad,' he said. 'We took a lot of trouble designing them so that they didn't interrupt your eye too much, as bars tend to do. Babies bite them, they teethe on them. Go on, Gamba, go down on the floor and walk like a brigadier general . . . Giles, Giles — look, he's got such a lovely Prince Regent hairstyle. Hello, Giles, me old cock.'

In the crowd a man sneezed. Durrell raised a mock-critical eyebrow. 'It's a good job they're healthy,' he said dryly. 'You'll have to travel a long way to see animals in this condition. Normally in zoos their bellies are right out here, rather like mine, because they sit about all day with nothing to climb on.'

The female orang thrust her hand through the bars, picked up some straw and waved it temptingly at Durrell. 'Look!' cried a woman, 'he's got his hand out, look!'

'Yes,' Durrell said. 'He's about as trustworthy as a politician.' More laughter. 'No, I don't want your bed, thank you very much, terribly decent of you; no, you're not going to grab me by the wrist. In a zoo I visited recently the orangs were in a cage half this size and absolutely bare, not even a shelf, and lit by a tiny skylight, and the male was devoting his time to picking up a small fragment of sacking and placing it on his head. Talk about cruelty! I sometimes wonder, you know, whether it's worth preserving animals for humans.'

'Surely that's not what you're preserving them for?' said one of the visiting party.

'No, I know,' Durrell said with a grin. 'But don't let on.'

They moved on to pause outside the newly constructed quarters of the bare-faced ibis. The enclosure was now double its original height with a rear wall of Jersey granite topped by nesting boxes, which had tempted these ungainly birds to reproduce. The species, which nowadays lived on the cliff above one Turkish village, was down to fifty pairs in the wild. They were further threatened by the locals who objected to living under a drizzle of birdlime, eggshells and fusty straw. 'Haven't you seen the situation there?' Durrell snapped with the anger that rose in him whenever humans were thriving at the expense of animals. 'It's futile, as it so often is, even to try to preserve them on the spot, I've never heard such balls in all my life, a waste of time, energy, money and everything else. The birds crap on the people below, the people tip rubbish into their nests, so they haven't a bloody hope in hell, it's ludicrous even to . . .' Words, as often in rage, failed him. He spluttered. At such moments he confronted the absolute impossibility of his task. An ibis flapped clumsily across the encaged space, drove its long curved beak into a morsel of meat and ascended heavily to a perch. Sunlight gleamed among the iridescent feathers of its head. A wily eye blinked once.

Another small crowd had gathered round the indoor cage that housed the baby gorillas. Two girls in jeans, with notebooks, toys and baubles, were squatting in the cage, intent upon the babies, who hopped, skipped and jumped round these substitute mothers, black round eyes innocent to the point of extreme mischief. Currently the evidence in this comparative study of baby gorillas and human infants suggested that the apes, for their age, were the brighter and more original of the two. Here again, as soon as Durrell appeared on the scene, after an awed moment of recognition, the public began to ply him with questions. How old were the babies? How long were they inside the mother?

'They're just a year old, they're both boys,' Durrell said, once again with an enthusiasm that implied he had only just discovered for himself this unique information. 'It takes the same time, nine months, and the mother looks just as disgruntled towards the end as any woman does. We had to take them away because one of them simply kept her baby at arm's length from her breast and the other is terribly women's lib and decided she just didn't want to have the bloody thing in the first place, so dropped it on the floor and left it. But from the point of view of the production belt this is ideal because as soon as we take the baby away they can become pregnant again (I told you this zoo was sex-mad), otherwise we have to wait two years. And I'm quite sure there's no psychological deprivation. Desmond Morris has a theory that a hand-reared ape won't know how to mate, and though it's perfectly true that they do learn by example, a chimp I reared in the house, like a child, knew exactly what it was all about when the time came, but couldn't, as it were, locate the hole – it just took time. In a family group the young gorillas watch Dad very closely and occasionally give him a helping hand. They take these matters much more sensibly than human beings do.'

At a lively pace the party moved round the Zoo. The amenities were spacious. A naturally interesting landscape had been adapted to the needs of the creatures inhabiting it. Meanwhile Durrell talked about the breeding statistics. In each department – lemurs, tamarins, pheasants, lions – a member of staff was summoned by Durrell to give a brief lecture on his pet subject. They were slightly nervous, as if in the presence of a headmaster. They were rather formal; several were trained biologists. They were wholly dedicated – the sine qua non of working here. They offered solid information, which Durrell interspersed with touches of humour while pretending to melt into the background. The effect of his asides was to humanise a scientific operation. It made it commonly understandable. He expressed surprise on learning the age of creatures he had collected and brought back to Jersey as long ago as the Fifties. 'They're going all grey and bald, just like me, they've even got the same sort of stomachs on them,' he said, patting his pet obsession. When it emerged from 'the endless fretful squeaking of statistics', as he called them, that twenty-odd hedgehog tenrecs had been 'churned out', to use his phrase, within a year, he observed that on Tuesday he had an appointment to see the Governor of Jersey, to discuss the protocol suitable to Princess Grace of Monaco's forthcoming visit to the Zoo. There was a disconcerted pause in the party. He raised an eyebrow. 'I'm going to suggest,' he said ponderously, as if comic effect were far from his mind, 'that we shift the entire population off Jersey and turn the island into a gigantic reserve, with gorillas on the north coast and orang-utans on the south. I see no reason why some new and doubtless more progressive civilisation shouldn't be established here, a fine example to the Western world.' The party tittered feebly. Durrell was almost serious, but had not quite held his audience.

149

At the rear of the deep field, several wallabies lolloped under a variety of eucalyptus trees planted a few years ago and growing apace. Here was a glimpse of the outback domesticated on home ground. And here, as Durrell's mind leaped from one seemingly insoluble problem to the next, it was the marsupials that were at risk. Of his delight in their existence and fear for their future he had written in *Two in the Bush*. 'The Australians quite naturally say, when I try to get a hundred thousand pounds out of them, what good is it going to do bloody Australia?' he said. 'They're right. The problem is too big. Everything's dying out, being killed, having its environment destroyed. And we're too small. Everything's living on here, but slowly, in small numbers. The only way of getting round it is to say that we are a training college – which is true. We're training people to send forth to the outposts of the Empire. But, like all the rest, we can't do it fast enough. We could have endless staff tomorrow at almost no cost – we get daily applications from very good, keen, dedicated people – but we just haven't got the accommodation for them and nor has Jersey. The cottage is stuffed to capacity, so is that wing in the manor, and even if you could find flats in the island they'd be too expensive for the kinds of practical idealist who want to work here on short commons. So what I want to do is build a block with twenty rooms, a television lounge, a flat for a caretaker, someone who will cook for them – and that's fifty thousand quid. But it means we could not only train people to send out to the far corners of the globe, but also accommodate visiting scientists who want to come here and study and work.'

A fussy flight of parakeets wheeled overhead and disappeared into the ivy-thick branches of a tall tree. 'What I'm trying to do in this collection,' Durrell said slowly, 'yes, I suppose I'd better try to tell this to somebody, spell it out, I always think people know what I've had in mind for five years and they turn out not to – what I'm really trying to do here is to build up groups of the different mammals and birds, so that when someone comes to this place they get a good, quick grounding in the entire universe. They look round and they get their minivision of the world as it ought to be, was, can and must be again. You don't have to house an enormous range of marsupials, but you have marsupials, and these people have to deal with them, as well as rodents, as well as primates – it's a nutshell we have here, and it has got to be shared, by more and more people, so that at least, scattered over the world, there are a few, a very few, who know what they are doing and know that on what they are doing depends the survival of life, of the life force itself . . .'

Durrell turned and guided his party politely back to the manor house. Inside there was suitable refreshment for the guests. If he found them specially to his liking they might be asked to stay longer, dine out, round off the evening. Otherwise, exhausted by a day which, like most Jersey days, had cost him a good deal of nervous effort, he would squat on the floor at Jacquie's feet and watch television for a couple of hours before taking a bath and going early to bed.

Friday

THE NEXT morning dawned wet and windy. 'Christ,' Durrell muttered, staring out of the kitchen window, 'I must be in Jersey.' He was scribbling some notes for his speech at the next day's Annual General Meeting. His handwriting looked cramped and awkward, as though a teacher had once forced him to write with the wrong hand. He had been putting off the task. Now, amid groans and glances at the weather, acid asides about the idiocy of zookeeping and the lost joys of the south of France, he would spend the day tinkering not only with the speech but with the arrangements. As usual the prospect of putting himself on public show appalled him. With a reluctance tinged with elation, he was brooding on how best to entertain the various notables – Fleur Cowles, Lord Craigton, possibly David Niven – who would be descending on Jersey to attend the ceremony. He settled to work with his customary pretence of hating every second of it. Within minutes his face would be pensive and serene.

To avoid being in the way I thought I would explore the island or those parts of it on which Durrell had impinged, and the people on whom he had relied, in the years since starting the Zoo. One of them was Hope Platt, widow of the Shell executive who had helped Durrell with the Zoo's finances in the difficult days. She lived not far from the Zoo – nothing was far in Jersey – in a fine white house lodged in a deep dell leading down to the sea. Cat on her lap, dog at her feet, she said, 'Gerry has no faults except intolerance of people.' Durrell had told me that she had helped the Zoo as much as anyone on the island. Indeed, in the early stages, it was she who had lured her late husband into launching the appeal that pulled the Zoo out of the red. 'I can't visualise life without Gerry, but I can't think why,' she said. 'He calls me a mountain of ignorance and I don't mind a bit. He says I look like a sea lion when I turn over in a bathing suit and I don't care a jot. With anyone else I'd be deeply insulted. But when you suddenly ring him up, if you're bored or lonely, and

155

say, "Can I come to dinner?", the answer is always yes without thinking. For any kind of help, moral, financial or just company, he never puts it off, he responds instantly. He's incredibly generous as a person. His virtues are very great indeed.

'Sometimes people don't understand what he's doing,' Hope said. 'Sometimes they don't listen to what he wants to tell them. I don't think he bothers about people he assumes aren't interesting. I think he perhaps should have to. He couldn't have better relations with his staff or his friends, but with acquaintances I think perhaps he doesn't take enough trouble to pour out a little extra bit of charm. The trouble is he's essentially honest, I think, and he wouldn't like to exert charm for its own sake or in a phoney fashion to get money from people. It's a virtue in itself, but self-destructive in what he's trying to do. He could easily change. Everyone knows people one has to waste an awful lot of time being nice to.'

Hope Platt had accompanied the Durrells on several trips, to Mexico, Corfu, southern France, the zoos of Europe. She too had noticed the exhaustion visibly lifting as the miles lengthened between him and Jersey. 'After a quiet hour he always began to sing,' she said, 'at first very quietly in Greek, and then it became more bawdy and more and more cheerful, and by that time the sun was out and he was himself again. The Zoo weighs on him. It's doubtful whether living over the shop is a good idea. But the alternative is being away for months at a time, which is also not a good idea.'

Over the years, as the strains increased and the committees proliferated, she had noticed changes in his disposition. 'He's getting more bossy,' she said. 'I suppose that's natural, he has something to be bossy about. But he is more difficult, more easily offended. It's a tremendous stress on anyone to be a one-man band. He has been so amazingly successful in working hard at the things he does that I hope he doesn't begin to think that he's the only person who can do them. That's what is really worrying me a bit. Also he's apt to close down, because he has been hurt by criticism or by lack of faith.

'I admire him very much indeed. I like to listen to him, I

even like to read him – some books very much, they're incredibly funny. In certain cases you feel he could have done better if he tried, as the school reports say. Everyone who reads him, his public's enormous, is utterly devoted. I meet quite a lot of Americans in the West Indies who say, "Oh, yes, Jersey – do you know the Durrells?" And then I'm in.'

That morning I saw other residents of Jersey – a businessman, a senator, a chartered accountant, all supporters of the Trust – who at once sprang to the defence of Durrell's charm, almost as though it were a fault. 'Gerry is not a social success,' said the businessman. 'He's too retiring, he has a health problem, he needs constant rest. But the island is socially orientated – once Gerry started accepting invitations he would be rushed off his feet. The reason why he is successful on the island is that he is not popular. He remains something of a mystery – which is why people visit the Zoo, hoping for a glimpse of him. He was asked on TV if it was true that he walked round the Zoo in pyjamas in the early morning to avoid the crowds, and he replied, "No, but now and then I wander round with my flies undone." People want to see him, but he's like an ape to them behind metaphorical bars.

'A tiny measure of his charm is that he uses swear words a lot, but in a way that scandalises no one. He really thinks of people – a Christmas card with a personal note or verse on it, a couple of bottles of champagne out of the blue for lending a hand. After a year of not meeting, we resume our talk where we left off last time. He has an orderly filing cabinet of a brain that remembers one's problems, the names of the children, the drinks one likes.

'All Jersey now feels it knows a bit more about conservation than most people. But Jacquie is worried about *his* environment, about conserving *him*. If not, he could end up overwhelmed by the biggest collection of bores going. I suppose fame makes it difficult to live with oneself – one of the penalties, oddly, seems to be that the thing Gerald Durrell most underestimates is Gerald Durrell. It seems that when he has achieved something, it's not entirely what he wanted – he's for ever looking

157

for the big chance, with nothing in it for himself. He has an expanding spirit, he's never half-hearted but has the strength of character to admit that he's not a social pigeon, he never asks for favours. I like him because he is all the things I want to be.'

Jersey's attitude to Durrell was at first suspicious, then impressed, later open-mouthed when a local group including Lord Jersey and Sir Giles Guthrie resigned dramatically from the board of trustees. Nowadays people left him to his own eccentric devices. He had arrived quietly with an immediate nucleus of support, even if most islanders could not believe at first that a zoo without tigers, giraffes or elephants was worthy of the name. What on earth was this Durrell (a Huguenot name, which helped his image) trying to do? On an island nine miles by five, inhabited by only 63,000 people, bush telegraph soon established that this outsider was up to something none too conventional. He also triggered public imagination, to the extent of parents taking their children to the Zoo, if not digging into their pockets for the cause. Teaching local opinion that in the long run the Zoo was as relevant to human as to animal life, indeed a blueprint for the survival of our own species, was taking time. Meanwhile many Jerseymen were admitting that cows were not the only creatures to contribute to the island's economy. Cruise ships from America put into St Helier four or five times a year. From every continent thousands of people sent money, adopted animals, paid visits when they could. In the eyes of the world, Jersey was now more than a tax haven.

One of the first people to spot the local advantage to be gained from Durrell's scheme was Senator Wilfred Krichefski, at that point in charge of island tourism. 'I claim to be the first person he came to see in Jersey,' he said, a kindly bespectacled figure of great acuity sitting at his desk in a shapeless suit. 'I was impressed, as one naturally is – how can't one respond to his particular means of arousing one's interest? Gerry is an individual born to be different – when men are becoming stereotyped into groups, one revels in it. He has the ability to create interest in people with only a passing knowledge of his subject, which he approaches in a humorous way that endears him and his cause to anyone with the sense to listen.

'With Lord Jersey and Sir Giles Guthrie I was one of the three founder trustees. I feel a deep disappointment in the result of local appeals. Animals? they say. But there are children starving and what about cancer research? That's the reaction of wealthy people, who will seek any excuse to remain wealthy. The States of Jersey, our government, recognises an obligation – hence our loan at a low rate of interest, which enabled the Trust to buy Les Augrès Manor. Tourism officially pays a thousand a year, as a gesture. Education helps in an indirect way, by getting the schools involved in the cause.'

The Senator paused thoughtfully. 'The Zoo, you see, has managed to obliterate normal commercialisation, which most people can't believe because most other people never do it. Even London gives the impression of a smelly zoo run for profit with cages designed for the public. Even Whipsnade for my money is terribly commercial. But here there's an atmosphere. It melts into the country in graceful surroundings, and it's cast in the same mould of informality as Gerry himself. Despite all the other pleasures of the island, the beaches and bathing which it was my job as president to vaunt, it's amazing the number of visitors who regard the Zoo as a must. It helps us all. Which is why I'm personally dispirited by the lack of practical individual support. With his inimitable style only Gerry would be able to make people understand. But somehow he can't and won't.'

In the early days one or two slightly unorthodox methods of fund-raising were tried on the island and that morning Lady Calthorpe, a Zoo stalwart, told me about them. She had joined that private menagerie of friends who secured Durrell's comforts and peace of mind on trips abroad, when in his customary fashion, as she put it, he 'brushed up reality' for his books. She sat now, handsomely prim in black, exuding affection for her former employer. He had persuaded her to head the fund-raising committee for six years between 1964 and 1970, a period of service that now provoked in retrospect the awed as well as fond exasperation which seemed the emotional norm among those who ever worked with and for Durrell. In one of his books he had plainly exaggerated their first meeting. 'To put the record straight,' she now said, 'he didn't put ten bob in my collecting box at Jersey Airport when I was flag-selling, but half a crown. It was for the Jersey Association of Youth and Friendship. But on the next flag-day years later Gerry typically arrived at the airport bearing champagne and a huge lunch, which brought the whole place to a halt and lots of extra money too – and then he did, it's true, give ten bob, and also volunteered me into doing exactly what he wanted, which was even more typical.'

In an effort to buy a mate for Pedro, the spectacled bear from the Andes, six hundred people were invited to a bear's breakfast party at which kedgeree, wine, spaghetti and coffee were served. Fewer than half turned up but the rest donated, and though the function raised only £90 it was odd enough to create a stir in the national press. The system applied to the annual Dodo Ball was to make the occasion seem exclusive by asking relatively few people, so that a waiting list clamouring for tickets formed in no time. Every year the ball had a different theme of decor and dress. For the Jungle Ball, Durrell himself scored a comic triumph by appearing clean-shaven, with a boot-polish face, unrecognisable in the flowing robes of the Fon of Bafut. On another occasion, in another mood, he declined to attend at all on the grounds that the ball coincided with Jacquie's birthday, which he wanted to celebrate quietly with friends.

Other fund-raising functions had to be exclusive for sheer lack of space. The pianist Peter Katin wrote to enquire how he could help the cause. He agreed to give a recital in a manor house with limited capacity but the finest piano on the island. Asked with tact if he would object to closed-circuit television for the benefit of the waiting list, he promptly volunteered to follow his recital, after a boiled egg and a glass of milk, with a repeat performance. Such examples of generosity with time and talent always produced an equally generous response in those with only money to offer.

I drove down to Catha Weller's cottage overlooking one of Jersey's small dramatic harbours. This ample and amiable woman, who had put heart into the Zoo's early efforts to balance the budget, was sitting comfortably with a strong drink on her window seat. Yachts tugged at their moorings in the fine clean light. A shellfish stall was open on the quay. Just retired, Catha was the first of the vital quintet of disciples – Mallinson, Hartley, Mallet and Betty Boizard, the young islander who oversaw the daily admin – to leave Durrell's service. And she missed it. 'I don't know whether he likes people,' she said, 'but he's very astute with them. He inspires the push that everyone needs, especially in an enterprise to which you have to give all or nothing. Both Gerry and Jacquie have an enormous appreciation of what's done for them and are never afraid to show it. I've never met anyone like them for making people part of the organisation, for drawing them into everything and so engendering enthusiasm. Nothing is kept secret, you're always being asked your opinion – you're enveloped! That's why the staff stay. It's why we're all still driving ourselves mad, just as obsessed with the bloody place as he is.'

Her blue eyes assumed a faraway look. Plump and sensible, Mrs Weller was also a necromancer. She had psychic powers that at the drop of a hint brought out cards or predictions. 'Only Gerry could interview people, as he did me, over drinks on a Sunday morning,' she said with acid humour. 'And as I walked in I recognised the room. I'll be seeing this again, I thought, so I must have got the job.' At that first encounter the Durrells had exchanged a nod, she was evidently at a glance just what they wanted – efficient, combative, motherly, intuitive, blunt and loyal. They plunged her at once into the accounts, at a time when the Zoo was almost knocking on the receiver's door. The funds could not even manage her salary, so Durrell personally paid her £20 a week.

'With adversity, as in war,' Catha said, 'one rises to occasions. There was a feeling of closeness because of threat. All five of us felt it, and in any case you can never sit back when Durrell's around. He gave me carte blanche to get the Zoo on its financial feet, having decided only to form a Trust after proving, if we could, our powers of survival. And though Gerry was pretty nervy even then, highly strung, his enthusiasm did the trick. And really all I did, from eight in the morning to seven at night (not to mention in nightmares), was to be firm on his behalf. I refused to allow any purchase unless we paid cash for it. This meant raising money at once, so we went round the pubs and cafés persuading people to chip in with two quid, which gave us a nucleus of members, from which we never looked back – I think there are over 15,000 now. I appealed to Gerry's fans, drawing on years-old letters about his books from kids who were by then grown-up, and got a 10 per cent response. But even now the Zoo can't support itself. A month's gate in the height of summer equals only the animal food bill for that month.'

Mrs Weller moved downstairs to the quay to buy some freshly caught crabs for lunch. At the Zoo, after finishing his notes for Saturday's speech, Durrell was lunching swiftly off his favourite tinned sausages amid mounds of mash from a packet, pretending not to miss the four or five poached eggs he would also have been eating if not on a diet. He had spent part of the morning in touch with airlines and other concerns which might offer him concessions for the trip to Mauritius. As usual before an expedition, every means of cutting costs was explored with care, even if it meant using lavatory paper stamped with a brand name or blatantly mentioning an airline by name in an otherwise poetic description of the jungle viewed from ten thousand feet.

Most of the arrangements were now in the hands of John Hartley, who reported progress every few minutes to Durrell, who in turn greeted every success with a mixture of triumph and incredulity. He was anticipating the trip with pleasure. Indeed, his expeditions had always started at the same pitch of excitement as that first glimpse of Corfu when Durrell was a boy. 'However well prepared, you're always swept off your feet,' he said that afternoon. 'I'd looked forward to that first collecting trip so long, so eagerly, that I could hardly believe it was really happening. I can't now. It's the same feeling of unreality as Alice climbing through the looking-glass. I remember approaching West Africa at five in the morning with the sun rising, all the mists being dragged up like skeins of wool from the trees. It had such a powerful impact that I was drugged by it for hours, even days afterwards. One glass of beer that morning and I was as high as an eagle. To sit there, drink a beer, watch a lizard, vivid orange and swimming-pool blue, just nodding his head on the balcony. It's there for ever in my mind, much more than reality, because it was alive and I was alive – and that has been true of everything, every trip. It's almost vulgar, the way I remember photographically in the colours of a glossy magazine.'

Hartley popped into the room with the news that Air France had granted a substantial reduction in the fares to Mauritius. Durrell clapped, then raised his eyebrows. 'Typically British,' he said in British tones, 'of British Airways. They've missed the boat again. Like Britain.' Then with scarcely a pause, though giving Hartley a fatherly nod of approval, he went on, 'For instance, the very first night at the rest-house in the Cameroons, after organising the lorries and seeing the right people – two essential activities in a country you intend to plunder (for all the right reasons, of course) – we had dinner and drifted down to the little botanical gardens there. The British always had a habit of making botanical gardens, like country clubs, wherever they went. And with torches we walked down a tiny stream with all this lush undergrowth. And like a couple of schoolkids we picked up whatever we found, tree frogs, woodlice, centipedes, anything, carted it all back to the rest-house in jars and boxes, and oohed and gooed over them all until three in the morning. And that has always been the tone – a sudden and unbelievable rebirth with every step you take.'

Durrell gave a deep sigh. Expeditions were also explorations, times when he felt and enjoyed his own sense of life echoing back to him from the dense jungle, the serenity of the pampas ('the closest approach to true peace I've ever known'), the proliferation of species in parts of the earth where man was by no means uppermost. He had never felt lonely or lost; there was always far too much work. Nor had he ever been dogged by a fear of the risks or dangers. His experience was that animals never posed a threat unless disturbed by human stupidity. You could always outwit them or take obvious precautions against injury. For the most part animals wanted only to be left alone, and although Durrell's purpose was to collect them, first for other

163

zoos, then in their own interest, what he most relished was observing their habits and habitat. 'As a naturalist,' he said, 'you have no idea, until you've experienced the tropical forest, how complex, astonishing and differentiated life is. Even in an English meadow you're blind to it. When I first read Darwin's outpourings in *The Voyage of the Beagle* I thought they were poetic licence – only to discover in Africa, and in Guiana, that he was grossly understating it. I've been trying to get it into my mind, not to mention getting it on paper, ever since.'

West Africa had come first, producing ('I haven't read it for years, thank God, so assume I can't have done justice to the place') *The Overloaded Ark*. Then, after a further trip to the Cameroons dominated as much by the Fon of Bafut's tipsy hospitality as by the sobrieties of wildlife, Durrell travelled westwards over the Atlantic to Guiana ('absolutely ravishing because totally different') to collect the creatures that prance and chitter through the pages of *Three Singles to Adventure*. 'In the Cameroons I was walking in a cathedral, staring endlessly upwards, only just able to glimpse the frescoes on the ceiling,' Durrell said. 'That was rainforest for me. But here it looked as though a florist's shop had been emptied over every tree or bush. I heard an odd noise, quite inexplicable, and was drawn to a tree rioting with orchids and giving out this susurration far more subtle than bees, and then with difficulty – for everything so fabulously new is experienced only with difficulty – I saw that something like two or three hundred hummingbirds were feeding off the orchids and making that intent perishable sound. And then on another occasion I was skimming in a canoe down the back tributaries and the whole surface was covered with water plants that had minute leaves rather like mustard-and-cress with a satiny haze of pink bloom hanging over it, so that miraculously your canoe was sliding without the faintest sound across a pink-flowered lawn. God,' Durrell said breathlessly, 'I must get back there, quick, quick, quick, before it's too late. It's getting too late everywhere.

'Will I ever wake up again as I did one morning in Georgetown? All so very apparently normal, like a suburban house in England, but built of wood, and not knowing where I was, expecting to spring awake and see the ship's cabin around me. But I looked out hazily, all I had to do lying on my pillow was open my eyes, and there was a huge magnolia tree with big glossy leaves like plates, all pullulating with three different species of tanager, an insect-eating bird with a soft bill, but red and black, cobalt, white and green, hovering in and out of the leaves as they fed. Such moments are beyond belief. And every expedition, twice to Argentina, in Malaysia, Australia, New Zealand, wherever I've gone, and now I hope and pray Mauritius, has left so many flashes in my mind of such absolute beauty, which I recall totally and at will, they're like touchstones for moments of gloom or despondency. They make me feel – if I may indulge in ponderous reflection for a moment – they make me feel that at least I'm at the centre of my own life. Being considered mad by everyone, mad and fanatical, suddenly makes sense.'

The afternoon moved on. Across the courtyard below Durrell's window, in still foggy weather, members of staff loped past on their various errands, grim, often bearded, dressed in practical student garb. The shriek of a bird cut the air. Monkeys quarrelled offstage. Visitors drifted past, dropping sandwich papers and drink cans, with the comatose and aimless air of people on holiday, waited to be animated by the next caged animal. Hartley materialised to give precise details of the timing of the Mauritius trip. Durrell's agent telephoned to ask with tact, amid a good deal of the coarse banter usual in Durrell's dealings with the literary scene, whether his publishers could expect a book about Mauritius and, if so, when. The author rang off, flopped into a chair, rolling his eyeballs. 'At least, when I die, there won't be any publishers in heaven – or in the other place, which they deserve even less for being so terribly good to me. Why don't I sit back and just write more about what I've done already? After all, I've got this total recall, even of word-for-word conversations – indeed, it's my only attribute. If something doesn't interest me I don't absorb it, it's like wiping a slate clean before anything's inscribed on it. But my memory is so exact that I have to go through it mentally with a pair of scissors to make sure I don't overwrite the bloody thing. I could easily produce another book on each trip, I might even be able to squeeze out a few last drops of immortal prose about the Fon. But somehow you can't, can you? That's why I keep having to go to different places. Eventually you get to the point of being incapable of another simile. There are only about twenty ways of describing an olive tree, and now I've used them all up I can't write about olive trees any more.

'I could actually redo Guiana,' he continued pensively, as though contemplating the notion for the first time. 'Because there was such an abundance of natural life. But the pampa is the pampa is the pampa — you have to rely on your companions as comic relief, if not as a source of metaphor, so to write again about it you'd have to travel with a completely new set of human beings. Australasia, alas, I've had to put in the deep-freeze compartment — it might suddenly pop out into honeyed words, but I doubt if I have the stomach for it — because if people on a trip are sulking around me in corners like primadonnas without a reputation, it just castrates me. Jacquie thrives on battle, some people do, they need the adrenalin. But I have to have a nice, calm, peaceful, friendly, happy atmosphere surrounded by people I love, if by people at all. Rows make me physically sick. They make me violent far beyond the content of the row.

'However,' Durrell added more placidly, 'I do consider my audience. Otherwise I need never go on a trip again, I could just sit in a hotel in Bournemouth and make it all up. Though I do think I'll write a good book about Mauritius, despite 90 per cent of the island being sugar cane. It sounds hell. But to go from villages where women wait for buses in vivid saris carrying dazzling umbrellas to lying face-down on the reef staring at fish, that's heaven. I thought I might buy a property there and spend my old age sitting in a wheelchair, if I live that long, and talk into a tape recorder — maliciously spilling all the beans I couldn't report at the time, the stuff we all insist on being absolutely hush-hush until you die. It could be valueless. It could also be fun.'

He stared out of the window as if at other continents. Outside the animals had begun dozing and visitors were hurrying off to fish suppers, seaside landladies, pubs. As usual in his own flat Durrell seemed not entirely at ease. Jersey was less a home for him than a sanctuary for the animals. He turned back from the gathering dusk. Mallinson was in the room describing an odd sickness that had struck a rare tamarin. Hartley was hovering for the day's last instructions. In companionship, on the note of festivity he liked his evenings to strike, Gerry asked Hartley to fetch a corkscrew and Mallinson to choose a bottle out of the dining-room cupboard.

Still brooding over his expeditions, Durrell said that visitors to the Zoo were always asking by name after particular animals he had collected in the early Fifties in the Cameroons, Guiana, the Argentine. He had described them in his books as if they were immortal. And at such challenges, he said, he often scratched his head and wondered where the hell these creatures were being kept. But of course only their descendants, if any, survived here in captivity. It was sometimes difficult to remember that the majority of animal lifespans were shorter than ours. The booming squirrels and flying mice from *The Bafut Beagles* were no more; indeed the Fon was dead too. The curassows and red howlers now existed only in the pages of *Three Singles to Adventure*. Durrell smacked his lips. He appeared to like harking back to these old events. The cork plopped out of the bottle.

'So time has passed.' He rolled his eyeballs. 'Happy days.' Whenever possible, he slipped in a despairing reference to his age: the Durrell fashion of touching wood. The 'happy days' in question were those of the early expeditions before he 'made the mistake' of starting the Zoo. In those years the responsibility was never irksome, as it was now, for ever asking where the next bananas or covenants were coming from. Three months saw an expedition through: the elation of setting foot in a new country, of learning the habits of animals he captured in a base camp which, like home in Jersey, was designed more for their comfort than his convenience, and of finally bringing them back to a future. Horizons out there were always relatively narrow. Life had an intense immediacy, to do with foodstuffs, security and sleep, how and where to find all three. Life was never usually like that. 'We are spoilt,' he said.

There was little doubt that Durrell had been deeply happy

167

during those expeditions, as he would be again in later ones. The trips were the essence of him: like everything else he took pleasure in doing, they were always adventurous, more or less single-handed, packed with short-term concentration, by definition dictatorial. He liked being a one-man band accompanied only by animals. The Durrell whom people met now in Jersey, however intently he put himself across when need be, was the shadow of his expeditions, rather than the spirit. On social occasions, which for him had no purpose, he was either humble or too assertive – often with jokes which never quite came off because they were expected of him. At any party not his own he was a fish out of water, a tiger deprived of the jungle. The trips reflected, with the clarity that made his books so enjoyable, his enthusiasm, his determination to do a job and do it well, his urgent sense of breaking into nature to serve it.

'I must say,' he said, rolling his tongue round a rather good Mâcon that Mallinson had picked, 'I'm looking forward to Mauritius.'

On this note, never less than surprising, Durrell reached across and turned on the television. Hartley put down his glass, waved and sidled off home to wife and child. Backing towards the door, Mallinson mumbled a farewell, fingered his tie and departed towards his wife and two children. Alone at last against an armchair, Durrell spread his limbs generously over the floor, gazed with a mixture of hostile fixity and indifference at the set and drank more wine, while waiting for Jacquie to call him into the kitchen for supper, unless they decided to have it on their knees.

Saturday

SHIRT-TAILS flapping outside the wrong trousers, Durrell stood on the landing of the flat looking disconcerted. Jacquie was insisting that for the AGM he wear the blue suit. 'I knew I was wrong to start this Trust,' he said, hobbling back to his bathroom where in ample cupboards a smaller selection of clothes than usual awaited him, for he had recently put on weight. 'I could save a lot of trouble and money,' he said, 'by just keeping the creatures I like and a maximum of three helpers. Or collect animals for bad zoos and winter in the south of France. Or just sit in the south of France . . .'

Picking out the blue suit, he put it on in grudging preparation for the ordeal of appearing in public. This annual Saturday was one of those occasions which Durrell most hated and best loved, when he was expected (and wanted) to put on a smooth and hilarious performance and also gathered round himself friends and sympathisers whom he wanted (and needed) to see and entertain. The AGM might seem a modest affair, of parochial size and character. Only a handful of local supporters would probably turn up, and from mainland or continent a few more or less notable names. But in Durrell's eyes the function had to vindicate, in the presence of however small a public, a year of intense private effort. It was a day to emphasise how vital the work was, preferably by being funny about it.

For an hour that morning, as Durrell struggled into the proper dress and nagged his lieutenants, there was the usual flurry of doubts, excuses and panics. Someone crucial to the event might cry off. A car had broken down. The last batch of incoming guests – Hartley being at this point on the phone to the airport – risked late arrival in Jersey, as the island weather was up to its usual tricks: bright sunlight on the beaches was often accompanied a few miles inland by fogbound runways. The Durrell credo had it that imminent disaster was the preface to success. Unless a lot of little things went wrong, no big thing could go right.

171

Jacquie stayed coolly detached from this chaos, knowing precisely what brand of support to give Durrell when required. Into the confusion now stepped the trim intellectual figure of Dr Thomas Lovejoy, grinning broadly, clapping the half-clad Durrell round the shoulder, dryly deprecating his cleverness in arriving on time from Philadelphia. In wholehearted welcome and an excruciating accent, Durrell cried, 'Hi there, li'l ole Tom!', clasping the man who had set up fund-raising for the Zoo across the Atlantic. Not only did he closely share Durrell's ideals but he also had the academic training to support them. Soon he would be escorting Durrell on yet another whistle-stop tour of wealthy mansions in the United States. Lovejoy was today's hero, the avatar of tomorrow's success. No extreme of wit or hospitality was too good for this much-needed friend in a part of the world Durrell had every intention of conquering.

It had not been easy to raise money in America or even arouse interest. According to Lovejoy, Durrell was himself one of the stumbling blocks. He put obstacles in the way of selling his wares or sharing his beliefs. People who attended his meetings – in Washington's Constitution Hall, for example, 3,000 had gathered – argued that anyone so funny about his work had no chance of achieving anything serious. In the second place anyone who drank so much (like, for instance, the Fon of Bafut) could do nothing responsible in life. Next, what the hell was captive breeding as a concept? Humour and alcohol apart, this last idea sounded to many ears like the experiments on humans in the concentration camps.

To counteract the prejudice, a scientific advisory board had been formed. It was studded with names. Durrell's work, to his relief, was seriously regarded by biologists of standing. But still the doubt stuck, making tougher than ever the exhausting process of persuading rich people to part with hard cash. Durrell had already whipped more than once round the moneybags of America. The papers had pounced on him as a newsworthy wag. Television had eaten him up with chat. Long lectures had consumed his energies. A few exclusive dinner parties had put him on his best behaviour. On all these occasions he talked at the top of his bent, projected film about the Zoo, sketched on paper while describing in words his adventures with species that mattered to the survival of the very people whose wealth he was trying to tap. But even where in theory people were as quick on the draw with a chequebook as with a gun, it was hard to establish 'the Durrell presence'. The disbelief was not in his work but in his capacity to do it. Jokes subtly told against him. His camaraderie over the champagne exposed him as no more than a witty guest.

As a matter of daily routine Lovejoy was always having to deal with crass questions. Even from well-informed persons, they displayed a basic ignorance of even more basic values. Why worry about endangered species? Surely species have always become extinct naturally? His answer, the only answer, was that until man came blundering on the scene, an 'orderly' replacement occurred in nature. A new animal developed to take the place of one less well adapted to a quietly changing environment. It was we who had accelerated the process, said Lovejoy. We killed without giving life. We took away without giving back. We had our own gigantically selfish priorities. Lovejoy pointed out that less than half the world's living creatures had even been named, let alone assigned their due importance in the scheme of things. It was still a mystery to biologists. How dared we interfere with a process – provisionally known as evolution – which was far beyond our understanding? It was like plunging a stiletto into our own brains in the fond hope, not even of revelation, but of crude advantage.

'The great thing about Jersey,' Lovejoy said, identifying, as we all did, the place with the man, 'is that it concentrates on nature as a whole. Some American zoos, it's true, breed the big, threatened, spectacular creatures. But no one else bothers about conserving the littler things, which aren't so beautiful unless you really look at them, which don't count unless you accept their mystery as a component in ecosystems, as genetic resources for man's use, and as an always equal but still unplumbed part of the evolutionary process to which man is subject and always will be. That's why Jersey is so vital. It works in intuitive leaps. It's a marvellous expression of Gerry's combination of clean science and inexplicable green-thumbery. It's a scientific institution that plods carefully on – nutritional research, breeding statistics and so forth – with an occasional injection of genius at just the right moment. And that's unique.'

Lovejoy looked out of the window, glad to be back. Cars had begun to roll up to the Zoo. The weather was lifting. Human voices called eagerly up and down stairways, echoing grunts and screeches from the park. Gerald Durrell, neatly suited, grave in demeanour, severe of eye, waited in a nervous trance on the landing to be shepherded to the vehicle that would lead him away to the place of execution. The mood was tense with not quite happy expectancy. The only relief was Jacquie. Plain-speaking and casual, she had decided not to attend the meeting. Here she was, a touch of the quotidian, making Durrell feel normal. She hung amiably about the top of the stairs, as if crisis happened too often to bother with, waiting to see everyone off before she got back to washing clothes, checking a car engine, writing cheques or pages for her latest book, *Intimate Relations*. It was no failure of loyalty. She was as keen on preserving his independence as her own. Life wasn't livable otherwise. She grinned, waved and whistled her way back into the flat.

The hotel where the meetings were usually held stood on a clifftop overlooking the miniature dramatics of the Jersey coastline. The sea below misted, brightly cleared, then fogged up every few moments as the morning changed course. 'Not every honeymoon isle can boast a Wuthering Heights,' Durrell muttered. With a preoccupied air, as if inwardly rehearsing his speech, he entered the severe stone building. In a long saloon, with picture windows turning the sea views into picture postcards, a muddle of old friends and people he half knew were settling into rows of upright chairs. They chatted, removing their coats, trying not to stare furtively at their host. Lord Craigton took his seat austerely in the background, awaiting his call to support the cause which he lost no opportunity of airing in the House of Lords. All smiles, Hope Platt seemed to be cordially anticipating a good deal of entertainment from the occasion or just from Durrell. One or two members of the Zoo staff lurked awkwardly in the wings, out of their element in best suits. Dr Lovejoy beamed steadily. A projector was in the process of assembly, leads snaking all over the floor. At the door a stall bearing leaflets, annual reports and souvenirs was manned by the office staff, looking bemused and smirking a bit. Several figures stalked about, not quite strutting, with an official look, whispering to one another. Moral support was in the air.

I could hardly look at Gerry. He seemed at his most vulnerable. His inner argument ran on simple lines. These people were expecting his all. There was no chance, he being human, fallible, weak and so on, that he could give it to them, however much the issues meant to him. This was as awful a moment as starting a new book, but at least that agony took place in private. He hissed rapidly in someone's ear. A very large brandy appeared in seconds. It vanished in a swallow. He waved aside the offered refill, ordering it to be available the instant his speech was over.

The hall settled back. As at all such meetings, one or two formal preliminaries stood in the way of entertainment. Brian Park as chairman uttered a few light, good-natured words, knowing quite well that we were all waiting for Durrell. Elections duly took place. Everyone on the committee was re-elected with the usual grunts from the belly and flutters of the hand from a well-disposed body proposing, seconding, as requested. The treasurer Robin Rumboll was now on his feet. 'The appropriate quorum being present,' he was saying, 'I would like on behalf of the Council to welcome everybody here. I am sure that you would like me to express your appreciation of, and indeed our admiration for, the Eleventh Annual Report. This is, I believe, a marvellous work, a collector's item, packed full of interesting, useful and valuable information about the species we are trying to preserve. I think we might be justly proud that we can produce a document of this standard . . . Now for the accounts,' added Rumboll to a flicker of all-round amusement, going on to show that even the balance sheet was in good heart. A vote was taken. Nobody objected. He laid down his accounts. 'Well, ladies and gentlemen,' he said shyly, 'I'm pleased to say that concludes the formal and somewhat drier part of this occasion and I would therefore like to declare the Annual General Meeting well and truly closed. And now may I introduce you to Gerry Durrell, who I think is going to keep you amused for a minute or two and report on the Trust's progress this year.'

Durrell stood up and coughed, looking severe, almost as if to emphasise his nice sense of timing. In the detested suit he wore a statesmanlike air, a little larger than life, very much in command of himself, a slight twinkle in the eye. There was no trace of nervous tension. His voice was clear, amicable and easy-going. He appeared to be addressing an audience whose sympathy he could take for granted, very much as he might in his books. He was straight-faced, but vigilantly on the lookout for a joke to alleviate the sobriety of his message, an anecdote to pinpoint the argument.

He began in low key by announcing his awards to various people at the Zoo for successful breeding – 'not among the staff'. The recipients of silver medals loped sidelong to the platform in their unaccustomed clothes, rather like animals out of their native habitats, shook the Honorary Director's hand, grinned at some sotto voce crack or veiled insult, and scuttled back into the audience. With verve Durrell then talked of two recently born gorillas, one of whom was the most beautiful thus far bred at the Zoo, 'despite the fact that when I was about to kiss it yesterday it tried to bite me – but then my wife does that so frequently that I get used to it'. Durrell's manner in speaking of gorillas was one of half-concealed delight, like a parent boasting about his own progeny. His paternal instincts were at the disposal of the endangered.

He went on informally vouchsafing details of the Zoo's success in breeding – Jamaican hutias, bare-faced ibises, the spectacled bear. As always, Durrell exploited lightness of tone to conceal the fact that he was quietly moved by these results. He wanted not to show emotion, merely to divulge the facts. But just because it was being shamefacedly hidden, the passion Durrell felt was all the more moving to his audience. He began to look even more vulnerable than usual, but without losing an inch of his authority. He began too to look more isolated. Standing alone on the makeshift platform in a sky-blue suit, capable of taking any weight on his broad shoulders if people only offered a feedback of sympathy and money in equal quantities, he could and would carry on the good work. Millionaires were helpful, yes, but no more so than a small boy who put his pocket money into an envelope and sent it trustfully to the Trust.

'As I say, we love the millionaires,' he said, 'but we love the small people as well.' He paused. He turned, to an excited murmur and a sprinkle of applause from those in the know, and confronted the easel behind him on the platform, to which was attached a block of drawing paper a yard wide. Taking up a thick-tipped felt pen, he faced the blank white sheet, coughed again and took his time. Most people were aware that Durrell's simple cartoons of animals, often catching their personalities in a single detail, echoed the quick effects of his prose. A few swift strokes produced vivid characters. As his hand moved easily across the paper, his back only half-turned to the audience politely to let them see his every line, he began chatting over his shoulder, pausing at each phrase, syncopating his narrative with yet another touch to the picture. 'The catching of animals – though it's difficult – is nothing like as hard – as the keeping of them,' he mused, as half a pert and gloomy small owl formed on the paper. 'And of course food is the most important – especially when – you haven't got it.' He sketched in a beak.

'In West Africa once – I was brought a female scops owl – and her babies. Well – ' He added an identifying feather or two to the mother. 'Yes, something like that. That's Mum. But of course the babies – don't look like that at all.' The audience tittered expectantly. 'They're white – and covered with fluff – and they look as though they've just come out of a rather unsuccessful session in a spin-dryer.' The nicely timed words matched the picture of the chicks, and everyone laughed. Almost too quickly he went on, 'At the time I hadn't access to all the delicious things owls so adore – all I had was raw meat – but, as you know, they eat everything and then in a terribly ladylike way they regurgitate all the feathers and bones in the form of a pellet.' He paused. 'So I wrapped the piece of meat in cotton wool.' He added a finishing

177

touch to the drawing. 'Which worked an absolute treat – except' (his voice hurrying the punchline) 'the mother looked as though she was having a rather irritable snowball fight with her babies.'

Amid laughter he tore the drawing off the pad and let it fall. It slid across the parquet. Within a quarter of an hour seven or eight others had joined it. Durrell disparaged his own talent for light entertainment, except for its power to draw funds for the Zoo. He saw himself as a doodler, a dab hand at greetings cards, a sketcher for any little occasion that needed a visual joke to celebrate it. His drawings were often given as compliments to friends or helpers – like all his presents, thoughtful, unexpected and individual.

He now addressed himself to the blank paper and began to trace some lines that might turn into anything. 'On another occasion,' he said, 'we got a baby screamer. Not at all like the parents, of course – a high, domed, intellectual forehead – ' (a plumpish philosopher began to take shape on the easel) ' – and no brain whatsoever. Covered in yellow fluff – and very tiny little wings like that – and the most enormous feet, so big they're apt to tread on their own toes and fall over.' By this time Durrell had so heightened the mood of his audience that he was prompting maximum laughter with a minimum of humour. 'And again this thing wouldn't eat – but – it did display a faint interest in spinach.' More laughter interrupted him. The risible bird on the paper seemed to be gazing with a warily greedy eye at a tedious diet. 'So I chopped up – and gave it – some spinach. And then,' he said with relish, 'I had an idea and said to my wife, "You know, I think we should do what the mother bird does." And she said – ' his voice grew ominous ' – "What does the mother bird do?" ' Marital discord loomed, and the audience loved it. 'And I said, "Well, she chews it all and brings it up again and gives it to the baby." So my wife said . . . "Rather you than me." ' The audience now in fits of giggles. 'But I said, "Well, I'm afraid it'll have to be you, darling, because I smoke." ' Further giggles. A pause, Durrell still quite deadpan, elongating the pause. 'So for the next month or two she chewed up platefuls of spinach three times a day.' Gales of laughter, Durrell still soberly awaiting his moment for the careful (and no doubt true) dénouement. 'Even now,' he said, 'it's not among her favourite vegetables.' Which brought the house down.

Rather than coming to some graceful conclusion, Durrell simply stopped, lumbered off, sat down. There was applause. A neat, authoritative man, whom nobody in the audience seemed to know, stepped through the clapping to collect all the drawings Durrell had tossed on the floor. While the artist himself slipped into the wings to claim his second brandy, this fellow stood crisply before the assembly and suggested that these rejects of Durrell's talent be instantly auctioned to provide funds for the animals. He had gauged precisely the spirit of the audience. They were all in generous heart. In a few exuberant moments, of which he took full advantage (he turned out to be an auctioneer by profession), he sold off the drawings at good prices. This made a brisk end to the speeches, preparing the way for the final act in the morning's entertainment: a showing of one of a recent series of half-hour films about the Zoo made for Canadian television. The film was said by Durrell, whose experience of the film industry usually aroused scorn and impatience, to be the most sensitive record so far attempted of his work.

The projector whirred. Long curtains were drawn to shut out the panorama of sea, cliff and wooded slope which had been changing mood all morning from hazy sun to a threat of rain. As the titles came up, the audience settled down to watch Durrell as his public saw him, talking to millions, trying to wring their hearts and empty their purses, with a very straight challenge not to their charity but to self-interest. At once a vivid image riveted the attention: fingers in huge close-up snuffing out the flame of a candle – so easy for us humans to extinguish a species by caring too little about the balances that surrounded us in nature, balances that enabled us to survive on a planet which we were busily engaged in turning against us. Pictures of threatened animals, pursuing their lives and habits within the freedom of cages, began to work on the hushed audience. Durrell's homely eloquence on the soundtrack did the rest.

His experience with films was wide. He had worked

several times, in Australia, Corfu and Sierra Leone, with Christopher Parsons, a producer with the BBC Natural History Unit at Bristol. All these expeditions had been dogged by the ill luck, acrimony and ego that seemed inseparable from the making of good programmes. You had to be tough, according to Parsons, in ways that Gerry never was or could be. The pressures of film-making, he said, were apt to produce an artifical Durrell. 'The humour goes sour on him under strain. The thought-out comic lines don't work, even on a commentary track, because all the fuss and bother makes him self-conscious.'

The worst instance was the filming of *Catch Me a Colobus* in Sierra Leone, an animal-collecting series out of which Durrell was supposed to cull a book. Indeed he wrote the book, but it was a muddled performance, reflecting only the bad spirit of the trip. The technicians refused overtime when a particular animal happened to be available for shooting. A cameraman was cleaning lenses when he should have been halfway up a tree. After waiting hours for a rare creature to run across a branch, someone in the unit coughed, frightening it away for another hour or two of overtime, not to mention fatigue or anger. 'Unit problems got on Gerry's nerves,' Parsons said. 'They worked at him like a sore, bit into him and incensed him, discolouring the whole trip. Perhaps because of it − I believe in such things − he hurt his back so seriously that Jacquie had to come hotfoot from South America. Fourteen weeks is a long time for a handful of people to live together in the isolation of the jungle, in particular when the leader quite reasonably expects everyone else to believe in the project to the exclusion of personal interest.'

Durrell had enjoyed similar bad luck with the filming of two of his most intimate subjects. *The Garden of the Gods* was about Corfu, an old paradise that had to be encapsulated for television, a task by definition impossible: Durrell had done it unrepeatably in *My Family*. The shooting of the film was marred by bad blood. One actor playing a Durrell brother refused to speak to another, ignoring him even when eating in the same restaurant, a far cry from crazy lunches at the daffodil-yellow villa. All this fuss meant that Durrell had to carry an emotional burden that had no point. The present capers of others were spoiling his past. At the time he was noticed drinking a bottle of ouzo before lunch every day. He had a breakdown in his once beloved Corfu. His sober conviction was, as he had said, that Corfu had surrendered its charms to beach hotels and package tours. No longer was heaven there. The place had become a hell on earth.

His beloved Jean Henri Fabre had also missed out. Here was the noblest naturalist of France, a local, his place not far from the *mazet* on the other side of the Rhône. His was the voice of the nineteenth century on insects. Smoothly combining the skills of the scientist and gifts of the poet in his lifelong observations, Fabre had fascinated Durrell's boyhood with his close-up look at insects and imbued him with a taste for the south. Durrell had wanted to film Fabre for the BBC, but insufficient money was forthcoming to make the all-round study that his love for the man demanded. Without funds, he insisted, he preferred to leave Fabre alone in his genius. But after much preliminary search by Durrell, someone else made the film on a low budget, therefore at a low key, thus making it impossible for years to do justice to the subject.

Such brushes with television had saddened Durrell. He admitted that he had perhaps thrown in his hand because he knew the monster was too big to fight, albeit too influential to ignore. 'We don't need TV,' Jacquie had often said defensively, perhaps fortified by the film options on certain books – *Rosy is My Relative, My Family* – which kept filtering into their bank accounts. But the pain seemed to go deeper. It now seemed less likely that Durrell would make a good film, despite his eye for it, before he was sure that he had absolute power over it. Even at lower levels, on a chat show like *Parkinson* for example, he now always asked either to be alone or to be told exactly who would be sharing the limelight. The publicity was not enough. It had to be of the right character – for no vain reason, only because the issues were too important to be buried, joked away by idiotic self-advertisers, hustled to the sidelines by too many people getting it all wrong.

This Canadian film drew to a close. As to content and cut, Durrell had insisted on the final word. It was an entertainment but also a picture of the inside of a man's mind, a summary of half a century of obstinate self-education, and a message. As people rose, stretching, to their feet, it was clear from their quietness that the message had gone home. No one sought Durrell out to shake his hand or place a substantial cheque in it. It was just as well. He was sinking another brandy at the bar and wondering what to have for lunch. In his usual mood after such gatherings, he was assuming the whole episode a failure and trusting it wasn't. He pouted dubiously through his beard whenever someone on the staff paid him a compliment. The audience drifted off.

The chosen – trustees, honoured staff, possible donors – climbed into a number of cars and drove through that potted landscape punctuated by cabbages and cows ('hardly Victorian England and partially not France,' as Gerry called Jersey) to the Moorings, a restaurant fronting the harbour of fishing boats at Gorey. Post-mortems and dozens of oysters launched the lunch on a tide of Chablis. We were Durrell's guests, the bill to be paid not by the Zoo but out of his own pocket. It was his pleasure, perhaps, apart from food and drink, his one true pleasure of the day. In his sharp blue eye there was a glint of liberation. One more meal, shared with friends, was saving him in the nick of time from the agony of confronting the ideal self he would like to become, the ideal world he wanted to create. The service was perfect, the wine delicious, the oysters local, the company to his taste, the strain over. What more could a man want?

'I've had a lucky life,' he murmured at some point during the lunch. 'In the sense that I've had everything I wanted given to me at the time I wanted it. That's not to say there aren't a lot more things I want, if I'm given the time . . . Everyone told me that starting a zoo was impossible, so I started my own zoo. Unlike Larry I've always lived with a bar of chocolate in my mouth. He had this terrible thing in him which he couldn't lance because nobody gave him a scalpel,' he added mysteriously. 'But I've had strawberries all the way. I feel guilty. I shouldn't carp at all, but I do because I'm human – or believe I am (or was). Oh dear, this sort of thing worries me about the human race, the fact that they don't *know* anything about themselves. They stick to their back gardens . . .'

Rolling splendidly out of the restaurant, Durrell blinked in the fading light of the week's last day. Like most weeks, it had been full, frustrating, busy, enjoyed, the agony counterpointing the pleasure. A sunset glimmered over the harbour, yellowing the white hulls of yachts askew on the ebb tide. He climbed into a car amid farewells from his guests and was driven back to the Zoo for an evening of recovery after the efforts of the day. Even the act of entertaining people stressed a man who gave so much to it and knew he was doing it to reward and to gain, as well as to show, affection. He dozed.

I thought back to the meeting. Several of the men wore discreet ties in dark blue decorated only with a small emblem under the knot. The same emblem appeared on certain cufflinks, in red on the cover of the Annual Report, on lapel buttons, and wherever any printed publicity for the Zoo was to be found. It was the dodo, symbol of the Trust: the dodo, that almost comic bird wiped out in Mauritius, its only known habitat, by the tragic ignorance of seventeenth-century man. Anyone who wore it was continuously aware of what the dodo stood for. It summarised in a flightless image the story of man's inhumanity to everyone else. A century later 60 million buffalo roamed the Midwest in their huge, inoffensive herds. In less than two hundred years the prairies were piled high with putrid carcases – Buffalo Bill's average bag was 250 a day – with no more than a few dozen left alive. The species was saved from extinction in the nick of time. That audience also knew that the passenger pigeon, a tasty bird of a trusting disposition, used to blanket the American skies in flocks 2 billion strong. Branches cracked under their weight as they roosted in forests miles long. As easy to shoot as to eat, 18,000 of these birds passed through the hands of one New York dealer in a single day. By 1914 one bird remained. It died in the Zoo at Cincinnati.

The deaths of those American species were caused by greed and sport, motives both acceptable if not pushed to extremes. But they always were. And the present was no better.

183

Governments in rivalry, or rather in concert, were steadily achieving the extermination of whales. Of the many species of seal, only a few had not been clubbed towards extinction. Eggs of rare birds were the more desirable to collectors as they grew rarer. The great auk vanished in 1850. The ostrich clung to our world only because women no longer thought feathers in fashion. A year before the First World War, 160 tons of feathers bedecked the women of France alone. The Indian lion, once spread generously across southern Asia, was now confined in small numbers to one tiny area by the colonial habit of bringing home a rug for your retirement. The couple of hundred left were now at the mercy of sheep and goats nibbling away their shrinking habitat.

I realised that everyone at the meeting had known these basic shots in the conservationist locker. That did not, for anyone, diminish their horror. The facts were impossible to get used to. And the list was endless. Durrell was always quoting them. He kept finding novel ways of presenting their plight. In desperation he anthropomorphised them. He stopped at nothing to wring our hearts, while well aware that too strong an appeal made generosity shrivel in a lot of people. He picked out phrases to make an animal our friend. Elephants were now plodding on their wastepaper-basket feet through the jungles in massively reduced numbers. Gnus with their flywhisk tails were on their way out. Just as the Alpine ibex had vanished for good because its body was superstitiously regarded as a medicine chest to cure all ills, so now the rhinoceros was on its last legs – ten years ago in Java only two dozen remained – because of the aphrodisiac reputation of its horn.

Meanwhile the Carolina parakeet departed this world in 1919, because it ate fruit meant for market. Several species of ungulate were done to death by Boer farmers, because they consumed the grass intended for, but not liked by, imported cattle. Recently half a million such hoofed creatures were systematically slaughtered in Africa for acting as host to sleeping sickness. Then someone discovered that birds were also hosts, as were small mammals. When a species became really rare, like the orang-utan now, the animal dealer moved in to poach it a step further towards the end. For every animal captured and sold to a zoo or laboratory, several died in escape or transit.

But there was a subtler spectre. The lesson was taught, though never learned, early in Western history. In antiquity the rich lands of Greece and Spain were eaten bare by grazing livestock and denuded of forest cover by shipbuilding. The soil eroded to rock.

The principle was clear. Once lost, land could never be regained. Only if bred in captivity might animals be returned eventually to their former strength. By now only a fifth of Madagascar was left for the lemurs. Everyone knew that the soil of tropical forest, cut down for agriculture, was speedily exhausted by first crops. We knew that ponds used as dumps wiped out the miniature world of toads and frogs drawn back by millions of years to breed in their birthplace. We saw oil damaging seabirds. We were aware that 70 untreated per cent of the excrement of Europe poured into lakes and rivers, on which in any case dams and weirs cut salmon off from spawning and denied eels access to the sea. In the Pacific, radiation had sterilised gulls. The radioactive particles, only half of which had reached earth ten years after a nuclear error or experiment, fell upon the bodies of animals and plants, bit into cells, roots, bones. One man in 1952 injected his garden rabbits with myxomatosis, directly causing, rather like the First World War, 10 million deaths in France and Germany alone.

The circle was variously vicious. A hawk ate a mouse which had eaten an insect which was poisoned. Or, more directly, if the insects were all dead the mice died of starvation, as did therefore the birds. In the West Indies men cleverly released the mongoose to suppress all snakes and rats, whereupon the rats more cleverly formed arboreal habits and preyed on whatever nested in the trees – at which, in turn, the starving mongoose rounded on man and savaged his chickens. I had felt the tension in that room at the AGM, everyone knowing there was no escape from the logic of human interference, which nonetheless spelled disaster.

Durrell woke from his snooze. In the twilight – Jersey's

temperamental weather had quietened into a cloudless evening – he stretched and toddled downstairs. He breathed the air, its animal must, a faint whiff of the tropics. Then, easing his shoulders, he began stumping about the premises with the air of a night watchman, as if to make sure no tapir escaped or marmoset was on the loose. But that wasn't it at all. Now and then he simply liked to check himself against the animals, the park, the place; to face what he was doing in the flesh; to prove to his satisfaction that all the words he had uttered during the day about the primacy and urgency of the venture, which depended only on him, were true. Or maybe it was again just his way of touching wood.

Almost to an animal, the Zoo was on the point of sleep. Lemurs lolled blinking on their branches at his intrusion. A lion responded moodily from the interior of an unseen lair. A few of the many white-eared pheasants – to which Jersey had given new life as a species – strutted and pecked at the grass over a last snack. Most brooded. The snowy owls stared back at Durrell unwinking. Rare parrots had their heads crooked under their wings, safe from extinction for another night. Though awake, the gorillas looked sluggish. A bored curiosity shone out of their eyes, sinisterly wise eyes in their shaggy caricature of a human face. They stared at Durrell, as though waiting for him to stop making fond noises at them.

Durrell turned away from his charges and went indoors to watch television. Soon, in his chair, one week ending, another on the way, one likely to be just as active, hopeful and funny, he was sound asleep, and snoring.

Epilogue

TWENTY YEARS later, at half past six on the evening of Monday 30 January 1995, my cousin Jennifer, who lives on Sark, rang me in London to say that Gerald Durrell had died that day. She had just seen the news on Channel Television.

He had done more than anyone I knew. He had probably enjoyed things more than anyone I knew. So there was no call to mourn him, but for that reason my eyes filled. Even more than Larry, so good at elevating a moment into an occasion, he brought a whiff of celebration to every encounter. It was always a party. Only Gerry would have insisted on a wedding that flew everyone in from everywhere and lasted five days. With acute envy of my earlier self — people dying quickens your relish for their being alive — I thought of the journeys up and down France with him, his wickedness, his joy in aimless talk, his roundabout way of pointing out to me that nature was cleverer than I thought, that appetite of his for things far beyond the comestible. I thought too of friendship, however you defined it — as love? I missed that link already. It had been an unfailing constant since the day he erupted in Soho Square forty years ago and asked me to come out with them all to Bertorelli's as if in the scale of things sharing a lunch had just as naturally important a place as starting a zoo.

After putting down the phone I walked upstairs to my room where several of Gerry's books were stacked at shoulder height on the shelves. I had read none of them for ages. But it so happened that I had bought that afternoon a disc of Ravel's piano music. I could not remember whether for Gerry such harmonies echoed France. They did for me. I put it on, recalling the winter mists of that place vaguely west of Paris, Lyons-la-Forêt, where we all spent a night in a creeper-clad hotel, scoffing and bibbing by firelight with the customary extravagance, and on a raw, damp morning I wandered down a lane before breakfast and gazed through high railings into the château where (I gathered from a plaque) Ravel as a guest had written *Le Tombeau de Coup-*

erin, to the plangent sorrows of which I was now listening. Each movement was dedicated to a friend of Ravel killed in the First World War, and the music sharpened the focus on old glimpses of my friend.

I had recently seen him in the liver unit at King's College Hospital. Opposite was a pub which overhung the station, Denmark Hill, where I had watched trains going southeast towards the continent. Thinking of France, I wished he would join me over a beer. At the hospital Gerry's ward was rattled by the same railway line. There, lying back, annoyed at being incontinent but on the whole good-humoured, he talked to me placidly of things wanted and missed. He described a moment at the Cromwell, his plush hospital in Kensington before the health insurance ran out, when, desperate for a return to overall wellbeing – a balance of sex, adventure, alcohol, luxury – he had insisted on drinking a single glass of champagne in a hot bath surrounded by his nurses. My throat felt constricted when I went away after that story; it had so much of Gerry in it, such desperation relieved by such hope and ending in pleasure, albeit brief.

Later that evening, Lee rang from Jersey and we were close to tears on the line. She talked to me of Gerry as if only now at his death could all his friends speak of love. Lee mentioned a service. She did not specify its nature. She assured me I would be invited.

At first I hardly noticed mourning. After a year of pain dying seemed the right thing, if not notably good. I was not up to imagining him in his coffin. Anyway it wasn't him, cut up in the postmortem or shrouded in the mortuary. He was elsewhere, but on the whole probably not here, at least not where I sat when digesting the news. He might visit me later. But wouldn't that be just imagination or wishful thinking? Apparitions were not chic these days. Gerry's belly would be a problem for a ghost. None of that prevented him turning up in dreams, of course. But wasn't I supposed to regard those phenomena as no serious form of reality? So what was I to do? The answer was to bite back the tears, tears more of frustration than grief, but it still felt like grief. I had no actual thoughts about Gerry's death. Feelings about it were still wrapped up in themselves. I regretted the departure of a beloved man of my own generation more sharply because he dragged me closer to death's door. Gerry's dying brought me within sight of the threshold.

A few days later Lee rang again from Jersey to ask us over for a ceremony in the courtyard of the manor. A stone was to be dedicated over Gerry's buried ashes. It now seemed natural that he had ceased to be, his death with luck a positive force for the future.

Travelling to Jersey was still small-scale. We were huddled together on a bumpy flight from Southampton. It felt like leaving a sophisticated continent for a scrap of land that looked only inches big from the air. The runway gaped below us like half the island. Time quickened as the plane throttled down. Within minutes we were whooshed through the system. Even more swiftly we hired a vehicle. Jersey's speed somehow matched its size. In this tiny car we were soon wriggling through a maze of only slightly less tiny lanes. At a remove of two decades I had remembered an undulating island with all sorts of secret places. Now it had turned into a suburb, the inland hills plastered with bungalows. On the coast far too many picture windows had sea views interrupted only by similar developments. Every inlet looked like a tax haven. Here was an island exempt from poverty, a paradise made squalid by wealth. Gerry was no longer there.

At Gorey we checked into our hotel next door to the Moorings, his favourite restaurant, which looked blank and out of sorts on this early spring afternoon before the start of the season. I had associated this little port, a castle surmounting it, with the utmost jollity. Now it struck me as supernaturally sad. Gerry had abandoned it. As yet we hadn't even set foot in the Zoo. Tomorrow's ceremony was a long way off. Even here where he liked to lunch I had no sense of him. The quay, the curve of frontage, had an abandoned look, and it was a relief no drive off through St Helier to the other end of the island to dine with Jeremy Mallinson, the Trust's Zoological Director, and his wife Odette.

Neither had changed, as I remembered them. In other

words, they had become more themselves: Jeremy whiter of hair, ruddier of visage, more assertive of view; Odette even more seductively eccentric than I recalled, which was saying a lot. From their house, which contained a fine collection of Lawrence Durrell apart from plenty of first editions by the younger master, we strolled down to a small harbour restaurant run by a Portuguese family, where I was delivered a pile of mussels which during the small hours asserted their evil. I was sick, I groaned, I lay back hopelessly in bed, I stared at the acidity of the dawn light, I knew I would never recover, and all this on the day when Gerry was due to be laid to rest, followed by a party which he would have much liked to attend.

I did not feel ill enough to resolve never to eat again. Indeed I was annoyed as violently as I was sick not to be celebrating him, probably at the Moorings, with just such a meal of countless oysters and large local soles, washed down by not a little Chablis, as once we might have shared. It was an unfair and unmerited sickness on such a day. Having slept not a wink all night, as liars always claimed, I had dry toast and thin tea sent up to our room overlooking the forlorn yachts, keels stuck in the mud. I had come to Jersey for nothing other than to acquire an illness that felt terminal.

All morning, wondering if I would make the ceremony, I stayed in that room, nausea drifting about it like an uneasy partner, while forcing myself to write in painful longhand a review of a comic novel which had to be dictated to the newspaper in London by lunchtime. I thought of Gerry's hand moving at witty speed over one foolscap page after another, as his hare easily outran my tortoise and, contrary to fable, won with disgusting ease. I remembered the occasion – a summer day that had seemed flawless until he played his trick – when on our arrival for lunch at the *mazet*, expecting as usual drinks to be instantly poured, he asked me to excuse him for a minute, he had some work to finish, called his secretary on to the terrace where she obediently opened her spiral notebook and poised her pencil. Gerry then solemnly dictated, 'The End'. He stared at me, blue eyes sharp with challenge. And we all burst out laughing, including the secretary, who promptly went off to type this final morning's work on the latest book.

It was Gerry's kind of tease, more than half fond but a good deal more than a quarter wicked. He knew I was having trouble with writer's block. The memory of that laughter returned this morning to mock a review that would cost a professional ten minutes. Yet, not without some soul-searching about my choice of career, I got it in on the dot.

Feeling not a lot better, I drove to the Zoo, watched my wife enjoy a nice-looking but to me repulsive lunch and grouchily pretended to enjoy gazing at one or two animals. Since I had last seen it in 1976 the place was of course transformed, unrecognisable, a complicated paradise of conservation. This restaurant, which gave visitors plenty of room to establish their own territory with children, parcels and talk, had views over the equally well-calculated habitats of other animals – but tucked away, amid foliage, behind banks of trees. Here the jungle had been shipped in, the extreme climes of the world had been naturalised in temperate Jersey. A meadow was a steppe, a cage thick with bushes the bush. It was hard, looking out over it, looking deep into it, not to think of it, today of all days, as a memorial to Gerald Durrell.

But such obvious thoughts were wayward. All good memorials, like this one, were an extension of life, indeed living organisms in themselves. Every single animal that had been brought here to breed, to form a society at human behest, the better later to accommodate itself to life on its own free terms, to return to its heritage, owed thanks to Durrell's persistence. He had created a country house where creatures in need were the guests. No snobbery was involved or class distinction. His method was to make nature pause, to impose a necessary moratorium on freedom for animals privileged by their rarity. By merely coming to the Zoo, thus supporting it, we visitors were part of the process of returning these creatures to life. Only to pay the entrance fee converted you into a conservationist. To come back time and again, as many did, turned you into a fanatic. These acts committed you. Here was a democracy of a curious sort: by paying a small fee you voted to be at one with the animals, you admitted you were in no way superior. You were saying, if only in a murmur, that we were all sharing the world with equal rights in it, spiders, accountants, poets, walruses, the man next door, goldfish, elephants, left-handed batsmen, flies and tycoons.

To me watching animals always seemed to clarify life, as watching human beings never necessarily did; we as a species were doomed to make war with one another, unless we found ways of referring back to nature. Durrell in Jersey gave us such ways, the revelation taking only a few hours with intervals for refreshment. Ill though I still felt that noon, I saw and admired this radical place for what it had become: a place to remind humans of their roots as well as to give other animals their continuance.

Weakened by all this reflection – 'thought,' as Gerry once put it, 'takes it out of you' – I drove back to the Dolphin in Gorey for a rest in the keen but gloomy hope of making it back to the Zoo in time for the ceremony at five o'clock. The sleep took effect. No nightmare brought me out in a sweat. Durrell missed his chance by not entering the fevered dreams I might have anticipated from last night's mussels. Was I sick because he was dead? Rubbish, I heard him say. I felt better at once.

We gathered at last in the forecourt of the manor on a chilly evening. I felt raw, distinctly rocky, but full of feeling, vulnerable to whatever might happen. Shouting against the wind, one or two people I remembered of old made valiant speeches about Gerry to dedicate the ashes buried beneath a plaque in the heart of the place he had created. The March evening did nothing to stem tears or protect the body from shivering or prevent me from thinking of Gerry with humour. The occasion bordered solemnity. We were all very quiet. I looked around, picking out faces I recognised or half identified from twenty years ago when I had been intimate with the Zoo in an earlier guise. A greying Robin Rumboll, the brisk young accountant who had helped pull the Zoo into profit; the grizzled Quentin Bloxam, now making a finely turned speech as Curator of Reptiles and Amphibians (when I remembered him as an inarticulate enthusiast swilling out cages); the trusty John Hartley looming over the occasion; in her wheelchair Catha Weller, who had once saved from bankruptcy a place that saved animals from extinction. As was inevitable, a number of absentees seemed present too. I thought of Gerry's mother who had ended her days happily in this last of her curious residences, of Larry who had died at home in the Languedoc at nearly eighty in 1990, of Jacquie who for a quarter of a century had kept Durrell company and now lived in France, and a host of others who had warmed to the cause. The story of Durrell's life stood around me in pensive silence. Now and then a screech from a treetop reminded me of the crucial point of that life. A chitter-chatter from a distant compound threatened the solemnity for a moment, then enhanced it.

The gathering round the plaque broke up vaguely, some people aiming off to cars, others winding up the staircase to the flat. Wine and food were there in plenty. Lee was resourceful about what in my condition I should drink, and with a wink we agreed on brandy, for its medicinal properties as much as for its associations. I had not seen Gerry's literary agent for ages; she also represented Larry. Talking to Anthea confirmed a shared but unspoken feeling that both her clients were still very much with us. So indeed, sitting in state across the room, was their splendid sister. Margaret had never looked any different to me from the time I first saw her in the 1950s. Nor did she now, surrounded by her sons, her grandchildren, with an air of matriarchal mischief. We hugged each other. She, this last surviving sibling of a family that had added immeasurably to my life and to the lives of countless others across the world, was as good as any of the Durrells at suggesting with style that on the whole life was timeless, altogether enjoyable, slightly beyond belief, a lot funnier than it might seem, as absurd as it was important, in the end a thoroughly good thing. And at last I began to feel myself again.